Dealing with virtually all aspects of scientific meetings, August Epple gives invaluable guidance for prospective organizers. He covers events from local afternoon Symposia to International Congresses with more than 1000 participants. He also provides insights for the tourist industry into the specific requirements that make scientific meetings different from others.

The author gets straight to the point, identifying common problems and offering solutions. In 20 chapters and an extensive appendix, attention is given to critical details such as: selection of the meeting site and timing of the event; stepwise program development; the selection of speakers and other key participants; social functions; budget matters; fund raising; the design of forms and brochures; publication of proceedings.

If you are organizing a scientific meeting this is your indispensable guide.

Organizing scientific meetings

Organizing scientific meetings

AUGUST EPPLE

Professor of Pathology, Anatomy and Cell Biology
Thomas Jefferson University, Philadelphia

CAMBRIDGE
UNIVERSITY PRESS

CAMBRIDGE UNIVERSITY PRESS
Cambridge, New York, Melbourne, Madrid, Cape Town, Singapore, São Paulo, Delhi

Cambridge University Press
The Edinburgh Building, Cambridge CB2 8RU, UK

Published in the United States of America by Cambridge University Press, New York

www.cambridge.org
Information on this title: www.cambridge.org/9780521563512

First published 1997

A catalogue record for this publication is available from the British Library

Library of Congress Cataloguing in Publication data
Epple, August.
    Organizing scientific meetings / August Epple.
        p.   cm.
    Includes index.
    ISBN 0 521 56351 8 (hardbound). – ISBN 0 521 58919 3 (pbk.)
    1. Science–Congresses–Planning.   2. Meetings–Planning.
    I. Title.
    Q101.E66   1997
    506.8'4–dc21   96-37969 CIP

ISBN 978-0-521-56351-2 hardback
ISBN 978-0-521-58919-2 paperback

Transferred to digital printing 2008

Contents

Preface

The unprecedented speed of recent scientific progress makes personal interactions between colleagues more important than ever. This is reflected in the growing number of meetings, ranging from small, highly specialized research conferences to large conventions with many facets, including a job market. Unfortunately though, many meetings are marred by major flaws that could have been avoided by expert advice. Of these, the most common ones stem from an underestimation of the necessary time and resources. Serious problems are almost certain when societies persuade inexperienced colleagues to run major meetings. In some cases, the participants will never learn of these problems. In other cases, tactful participants will not tell the organizer about them. Thus, a tradition of mistakes is often perpetuated.

This book cannot guarantee successful meetings. However, it should help to reduce avoidable problems. The author has had many years to make mistakes, and he has availed himself of this opportunity, as will be obvious from personal experiences given throughout the text. He apologizes for this lack of modesty. On the other hand, he often refrained from rubbing salt into the wounds of colleagues, tempting though it was. May this kindness induce them to buy, and not just borrow, this useful publication.

Philadelphia, Pennsylvania　　　　　　　　　　　　　　　　AUGUST EPPLE
Fall 1995

Acknowledgments

The author is indebted to many colleagues and friends for advice and encouragement. Special thanks go to Mary Adams-Wiley for valuable suggestions. Without the expert advice of the author's long-term collaborator, Barbara Nibbio, this book may never have been completed. Last but not least, it is a pleasure to acknowledge the efficient cooperation of Dr Simon Mitton and his staff at Cambridge University Press.

1

Introduction: what can go wrong?

Among the free lessons life has to offer, the mistakes of our fellow men are perhaps the best ones. The following tale may make the point. Let's pretend it is a figment of imagination.

A large society met happily, year after year, at their annual convention. There were many short talks each lasting twelve minutes, followed by three minutes of discussion. Every member was eligible to give a talk. While all seemed smooth sailing, some people began worrying about the quality of the presentations. They soon came up with a splendid idea that was immediately implemented: abstracts had to be submitted to committees which selected papers for platform presentation. Abstracts considered unworthy of the highest honors were just printed; no talk was allowed. The efficient secretary of the society mailed the decision on a postcard to all concerned. In this way, he saved the expenses for regular letters, and the members did not have to open envelopes.

As the mail arrived, there was happiness in some quarters. In others, weeping and gnashing of teeth; especially where technicians handed the mail to their masters with comments like 'Your talk has been rejected.' Devastated by the bad tidings as only sensitive scientists can be, some members went no more to the society's meetings; others, of a more flexible mind, went to pay homage to superior colleagues.

And what a surprise it was! One rejectee, a professor by rank, went to a session where he would have loved to give a talk. He entered the large room while it was dark; only the slide projections were visible. Soon he realized he couldn't grasp what the speaker, a graduate student, was explaining. Our friend was too shy to admit his ignorance by flight, and stayed. As the lights went on for the discussion, he noticed that the magnificent room was almost empty. Except for a few stragglers, only coworkers and friends of the speaker were present. The session was chaired by the speaker's graduate advisor and a colleague, the mighty lord of another research group. Depressed and humbled, our rejectee left the room.

Two hours later, he returned. The scene had changed. The two chairmen were still the same, but the small audience was different. It was now coworkers and friends of the other lablord; mainly graduate students, and almost all scheduled speakers.

Understandably, the experience left the rejectee with a severe trauma. From this, he only recuperated when he learned that the two chairmen had been in charge of the reviews of the session's abstracts.

We can learn from this episode. However, this memorable congress can teach us more lessons. The meeting had been convened in an attractive city during the tourist season. Result: (1) The room rates, even at the congress hotel, were too high; many members of the society did not come at all, while others stayed for only a short time. (2) The reasonably priced local restaurants were overcrowded, especially at lunch time; this meant that the participants had to choose between very expensive meals, or fast food in unpleasant neighborhoods.

The congress hotel had meeting rooms of every size and could accommodate all events spacewise. There was only one snag: the moveable partitions between the rooms were poorly insulated, and parallel sessions interfered with each other acoustically. Of course, nothing could be done when this was discovered on the first day of the meeting.

The organization of the social events was topsy-turvy. During the informal beer-and-snacks socializer at the start of the meeting, a magnificent big band played ballroom music. Almost nobody danced since there was not enough space, and people were not prepared. The band in festive tuxedos and the crowd, many in jeans and sneakers, just did not jibe. A few days later, the official banquet was held in a ballroom. The high price had discouraged most members from attending. It did not help that, hours before the banquet, red-faced members of the society's leadership ran up and down the hallways trying to sell tickets. And so, a small crowd was lost in a large room. The mood was dull and did not improve when a combo with a limited repertoire tried to play dance music. Mediocre food and expensive wines (to be ordered and paid extra for at the table) did not lift the spirits, either. To top it off, the leading figures of the society sat on a stage, from where they dispensed after-dinner admiration to each other and the members of their clans. The other participants could choose between feeling insulted or bored. Of course, most people fled the scene early.

Well then, what can go wrong with scientific meetings? Obviously, a lot. Unfortunately, however, many mistakes are repeated over and over again.

2

The decision: to run or not to run

2.1 Are you sure?

2.1.1 Experience and commitment

The preceding examples suggest that the prospective organizer of a scientific meeting had better look before leaping. To run a major meeting without prior experience would be foolish, no matter how much your friends encourage you. If you hope to muddle through, you are inviting disaster. Remember: the reputation from an unsuccessful meeting may stick with you for a long time.

For the novice, there are several ways to gain experience. One is to assist in the preparation of a major convention. Just carrying a minor responsibility, watching the progress of the preparations, and learning how unexpected problems are handled provides invaluable insights.

Another way to start out is to organize a special session, or a small symposium for a major meeting. Dealing with six speakers from the first letter of invitation to the receipt of the last manuscript is a good introduction to handling different, and more likely difficult, personalities.

You can also try it the harder way. The first scientific meeting I organized was a regional conference with more than a hundred participants. I had no experience, and major problems arose during the first morning: (1) The registration desk had been set up too late; some participants never returned to pay their fees. (2) Students and some of the faculty of the host institution appeared unexpectedly, refused to pay fees for an event in 'their' lecture room, and caused overcrowding. (3) The service in charge of coffee and cookies appeared when the morning break was over, almost at lunch time. (4) All the speakers had promised to hand in their manuscripts during the meeting, but almost none of them did. (5) Participants had to wait a long time for lunch because the recommended restaurants had not been alerted to the onslaught.

2.1.2 Purpose and quality of the meeting

Before accepting the responsibility for a meeting, it pays to analyze what can be achieved under the given conditions. A *regional conference* is worth the effort if it fosters cooperation and an *esprit de corps* among local scientists, even if the presentations are less than earthshaking. Besides, you can add Plenary Lectures by one or two 'hotshots' from outside the area.

The emphasis is different when it comes to national or international meetings. Of course, it is desirable that a major meeting also fosters cooperation and personal ties. However, your main concern must be the scientific quality, unless it is a reunion of 'old boys' of either sex, known as *'wandering circus.'* The latter type of meeting has its own momentum, with lectures of heterogeneous quality. If you can't get out of hosting it, prevent proceedings that could hurt your reputation.

A high quality meeting requires considerable diplomatic skills, such as: (1) persuading outstanding scientists to participate; and (2) keeping others from pouting because they are not giving Plenary Lectures. At this stage of your career, are you ready for the task?

The preparations for a meeting will vary with its scope: a conference with a narrow focus requires less time and effort than a more general meeting of equal size; and the larger the audience you wish to attract, the greater the need for a wider range of topics. This sounds simple, but is often ignored. A typical ill-conceived meeting will cover too few topics, leaving the majority of the audience disappointed.

2.1.3 Responsibilities and rewards

No matter what kind of meeting you are organizing, *you* will be held responsible for its success or lack thereof. Since you do not wish to be associated with failure, do not accept the job of an organizer unless you have strong input into the program. Do *not* get involved if committees determine the scientific program and, eventually, surprise you with their decisions. A common result of this system is a mess, with the schedule, list of speakers, and budget in limbo up to the last moment. If your authority does not match your responsibilities, don't take the job. Also note that the role of the organizer of a meeting should not be confused with that of the chairperson of a local committee (see Section 10.2).

If all goes well with a regional conference, your main reward may be the satisfaction of being a successful host; however, the chances are that your meeting will be forgotten soon. For the organizer of a big convention, the rewards can be limited, too. Often, he or she remains unknown to most participants, and the efforts are appreciated only by a few.

If you wish to erect a lasting monument for your ego, you must run a scientific top event with published proceedings. Naturally, it is easier to handle the proceedings of a smaller, specialized conference than those of a large, heterogeneous meeting. In either case, though, you will need an agreement with a good publisher; and you may not find one unless your proceedings have a reasonable chance of financial success. For details, see Chapter 8.

This may be the place for a serious warning: some people organize international meetings to make personal profit. This becomes obvious when participants realize that there is a difference between the prices charged for meals at the venue and neighboring restaurants. The best indicator of this kind of foul play is an outrageous congress fee that does not relate to the overall level of local prices. Never ever allow anyone to use your name as 'co-organizer' (or similar token official) unless you are sure you are dealing with decent people.

2.2 Can you do it?

If you have sufficient funds to hire a professional organizer, or you receive help from experienced staff of a society, you will be relieved of most of the preparations for non-scientific activities. However, *you* will still be blamed for everything that goes wrong, scientific and otherwise. Therefore, before committing yourself, more sobering thoughts are in order.

2.2.1 Money

Will you have enough funds for the following *expenses which will begin right from the moment you agree to run the meeting*?

(a) Telecommunications such as mailings, telephone, faxes, telex, electronic communications.
(b) Office supplies.
(c) Printing (e.g., paperheads and logo; preliminary announcements).
(d) Clerical support and equipment (especially word processing and copying).
(e) Advance payment for a professional organizer, in case you hire one and he requests it.
(f) A deposit to reserve a meeting facility (if you consider making the mistake of renting one; see Section 13.2).

If this list does not deter you, try to come up with a *realistic* budget estimate, as outlined in Section 13.4.

2.2.2 Time

Will you really have enough time, or are you underestimating your ongoing commitments to research, teaching, grant writing, administration, patient care and/or committee work? Take a moment and figure out how much time you spend already on your work. Then ask yourself if you can afford more professional activities, and how these will affect your private life.

If you are a junior faculty member, don't jeopardize your career by neglecting research and grant activities. If you are a departmental chairperson or dean, don't expect to save much time by delegating responsibilities. You will have to make many decisions personally, after thorough considerations. Moreover, do you really want to deal with more committees?

2.2.3 Local conditions and resources

The following list summarizes points for your consideration. For details on these topics, see Chapters 6 and 7.

(a) Location. Can your meeting place be reached easily by car, bus, airplane, or whatever type of transportation is required?

(b) Time. Is the envisioned date of the meeting realistic, considering local weather and traffic conditions? Think about seasonal hurricanes, snowbound airports, overcrowded highways and the impact of public holidays.

(c) Facilities. Will there be enough meeting rooms, and will they be of the right size? Will the accommodation be sufficient and acceptable for the expected participants? Will the local cafeterias and restaurants be open, and able to handle the participants?

On a campus, you may need the cooperation of the person(s) in charge of the facilities. In German universities, the good will of the superintendent can be so critical that an experienced convener devoted a book section to the handling of the 'Hausmeister' (Neuhoff, V. *Der Kongress*. VCH Verlagsgesellschaft mbH, Weinheim; 1986).

If your meeting follows an established pattern, and the regular crowd expects to be invited as usual, never ever select a site that requires restriction of the audience. In scientific circles, this could come close to a death wish.

(d) Equipment. Will the necessary equipment be available? Will it be free of charge? If you have to rent equipment, will the price be reasonable, and the available pieces what you really need? The devil is in the minor details: the most common problems are missing supports for projectors and screens that are too small.

(e) Staff. Will there be enough assistants when you need them? If you are counting on students to help you: will they really be available or will they be studying for their finals; or be off campus for their vacation? Estimate how many assistants you will need for the registration desk, slide projections and trouble shooting. Could you pay them if you can't get enough volunteers?

(f) Price. Will the local prices be acceptable to your participants?

(g) Safety. Is the neighborhood of your meeting place safe, or will you have to anticipate thefts, robberies, panhandling and other unpleasant situations?

2.2.4 Sponsorships

Who is going to sponsor your event? Will your institution, the city or the congress hotel provide meeting facilities free of charge? Who will pay for portfolios and their usual contents? Who will invite the participants to free receptions, concerts and other events that are expected of your type of meeting? For more on these issues, see Chapters 14 and 15.

2.3 Should you do it?

You could ask more succinctly: What will I get out of it? Obviously, the answer will depend on the circumstances and your personality. However, the weaker your professional standing, the more carefully you should consider this issue. Often, the organizer of a scientific meeting is in a no-win position. For every person he invites for a special role somebody else seems to be insulted.

When we held a regional conference in Philadelphia, Pennsylvania, people as far away as Arizona felt slighted. Why? We had invited a colleague from Washington State who happened to be in our region at the time of the meeting. Naturally, the friends in Arizona thought that we had brought him over from Seattle, all expenses paid. And what is good for Seattle is good for Tucson, Phoenix, and any other location. Right?

Take another not so hypothetical situation: How should an organizer, an assistant professor, act when approached by a bigwig, say a member of a body that decides on grants-in-aid for scientific research (e.g., an NIH Study Section or similar body) who wishes to give a talk but has nothing to say. Should the organizer decline the request, knowing that the fellow could swing the decision on his pending grant application − the application whose funding may decide on his promotion to a tenured position?

Let's modify the theme a little: A junior faculty member prepares a symposium, and his department chairman interferes though he has no right to do so. How should the young fellow behave? Obviously, like the hedgehog when he meets his beloved: with great caution. Of course, the hedgehog can buzz off when his skills of persuasion fail. Not so our assistant professor, and too late, it may dawn on him that he should not have organized a meeting before he was out of bondage.

More food for thought: A young scientist is the editor of the proceedings of a symposium. What clout does he have when it comes to the almost inevitable problems with manuscript delinquents? Take a typical case: the deadline for submission has passed long ago, but the proceedings have not been sent to the publisher because important contributions are still missing. People who handed in their manu-scripts on time begin asking when their papers will come out. As they learn of the delay, they get angry at the editor who does not dare to name the true culprits. Increasingly, the young editor's life becomes hell because the delinquents let him twist in the wind, convinced that they can delay another couple of weeks. Finally, he can't take it any longer and submits the available manuscripts to the publisher. Result: (1) The proceedings appear, dated one year later than announced. (2) Some colleagues call the editor incompetent. (3) The delinquents call him ruthless since he did not wait 'a few more days.' (4) Recipients of the proceedings are disappointed because expected contributions are missing.

When it comes to certain authors, even an editor with long-standing experience may have to cut his losses. However, he may be able to retrieve manuscripts at an earlier date than a younger colleague; and probably, he would not have invited habitual manuscript delinquents, regardless of their 'political' influence.

In summary: the decision to organize a meeting should be based on both its scientific merit and the prospective organizer's resources and clout. The bigger the meeting, the more likely are unfriendly encounters with lasting consequences.

3

Scientific and related events: variety delights

Most scientific meetings involve more than one type of event. One can look at these events as building blocks, and it is their combination and placement that will be instrumental for the success of the meeting (see Chapter 5). The following listing is arbitrary and merely highlights options that can be modified to suit a particular situation.

3.1 Scientific events

3.1.1 Lectures

All good lectures have one thing in common: they are not too long. It seems that all over the world the attention span has shrunk during recent decades. Blame it on our hectic lifestyle or the impact of the mass media: most people get restless when lectures exceed one hour.

To prevent monotony, lectures (with the possible exception of Main Lectures) should be followed by a discussion period and a break. The length of the break must vary with the circumstances, as detailed in the following sections. However, even 'Short Communications' should be scheduled at least three minutes apart. During sessions with several consecutive lectures, one or two extended breaks are definitely indicated (see below).

Within a series of lectures, it is imperative to *leave the time slot unused if a speaker does not show up*, unless the change can be announced well in advance. Otherwise, participants may miss the event. Any change of schedule during a meeting is likely to cause confusion.

Punctuality of lectures is a must when parallel sessions are held. A cautionary example: At an international conference convened in a country known for its 'relaxed' lifestyle, the projectionists appeared routinely late after lunch and resumed their jobs at different times. Nobody informed the gracious but overburdened organizer. But every afternoon overlapping talks caused anger, and foreign participants commented on the mess long after the meeting.

If you anticipate potential problems, try to solve them *before* your meeting, especially when you have to deal with labor union regulations. Otherwise, the following could happen again.

A major convention with parallel sessions was held at a university with unionized labor. Because of an idiotic contract, projectors and equipment were set up once a day, early in the morning. And that was it; no washing of blackboards, no new cups of water for speakers, etc., thereafter. Aware of the situation, I checked my meeting room, half an hour before the first afternoon session. Pointer and microphone were missing, and so I 'borrowed' them from across the hallway while everybody was at lunch. What if the projector had been missing, too? I don't even want to think about it. At any rate, my conscience was half clean after washing the blackboard of my meeting room. I considered further penitence by cleaning the blackboard across the hallway, too, but decided against it. If caught in the act, it could have been construed as an admission of guilt.

Let's now look at the different types of lectures.

3.1.1.1 Plenary or Main Lectures
Plenary or Main Lectures tend to be more formal than other presentations, and it is customary not to interrupt the speaker with questions. Usually, they last no more than sixty minutes, including discussion, and they come in two varieties: (1) Named Lectures, and (2) Unnamed Lectures. This distinction can be a major factor in the design of a meeting.

Named Lectures may honor an outstanding scientist, or acknowledge a sponsor. Because of the increasing number of honorees, several societies must soon run out of time slots. Sometimes, the dilemma can be resolved by scheduling certain Named Lectures only for alternate meetings. However, this may not be feasible if a society meets at four-year intervals. Another problem arises when Named Lectures are created without financial backing. Often, they are forced on a society during a business meeting when someone proposes to name a lecture after a person in the audience. Of course, the following vote will be positive. Only the organizer of the next meeting may cry in silence. Why? Because he has just been sentenced to raise another thousand, or more, dollars in travel funds. And it does not cheer him up either that the scope of the lecture will make the selection of a speaker difficult.

There can be more problems with Named Lectures: most notably, with the scheduling. Where do you find a time slot for the new lecture, honoring Sarah Stork, that is of equal status to that of the traditional John Frog lecture? You'd hate to open your congress with the Frog Lecture since this time the topic is of little interest to the audience; and besides, the presenter they forced upon you, is lousy. However, it would be sacrilegious not to begin with the Frog Lecture. So, while you know that the meeting will have a poor start, you rack your brains for where to place the brilliant speaker who will give the Stork Lecture; the lecture that would have been a smashing introduction to the congress.

For the organizer, the preceding considerations reiterate a warning: If you depend on other people's choice of the plenary speakers, you cannot develop the program beyond a rough outline until you have their decisions. The chances are that this will be late, and not without unpleasant surprises. If a society has Named Lectures without financial backing, think thrice: (1) You may have to raise funds for speakers, possibly for all speakers. (2) You may have to make financial commitments to the speakers before you have sufficient, if any, funds. (3) You may also be expected to come up with travel support for those in whose honor the lectures have been named.

Plenary Lectures should live up to their name and be of interest to the plenum. They should give a sovereign overview of a field to an audience that may not be familiar with details. Probably, it is impossible to draw a clear line. In general, Plenary Lectures should have a wider scope than State-of-the-Art Lectures (see below). Since the Plenary Lectures are critical for the success of a meeting, they should be given by true scholars, not data-makers with a narrow focus.

The best time slots for Plenary Lectures are the first ones in the morning and in the afternoon. Thus, Plenary Lectures may serve as introductions to subsequent sessions. In general, two Plenary Lectures per day are enough, but the glut of Named Lectures and 'political' considerations make this restriction often difficult. If necessary, Plenary Lectures can be scheduled after dinner. This is not an attractive solution since evenings should be used for interactive events, such as workshops, or the social program. Note: Plenary Lectures after dinner must not be confused with After-Dinner Talks (see below).

3.1.1.2 Seminar-Style Lectures

During Seminar-Style Lectures, the audience is encouraged to ask questions. However, even a stimulating seminar should not exceed 75 minutes, and the presenter should plan on a maximum of 50 minutes of personal speaking time. Lectures of this type may be most appropriate for research conferences or minisymposia.

3.1.1.3 State-of-the-Art Lectures

State-of-the-Art Lectures report on the cutting edge of science: the more recent the data, the better. This type of presentation is an excellent chance for young scientists to make themselves known, and to receive input on their work.

In general, State-of-the-Art Lectures will last 25–35 minutes, followed by at least five minutes of discussion. As with Short Communications, sessions with State-of-the-Art Lectures can be held in parallel. In the latter case, the presentations should be scheduled five minutes apart so that people can move from one room to another. If there are four or five lectures in a row, a fifteen-minute coffee break should be scheduled after the second or third lecture, respectively; and if there are six, a second break is almost a must after the fourth lecture. However, an ideal morning has no more than five consecutive State-of-the-Art Lectures, or four following a Plenary Lecture. If possible, no more than three consecutive State-of-the-Art Lectures should be scheduled for an afternoon, so that the remaining time can be used for events with more action, such as Poster Sessions, Colloquia or Workshops.

3.1.1.4 Short Communications

Often called 'Contributed Papers,' Short Communications are still a widespread, traditional form of presentation. Before the advent of Poster Sessions, they gave a maximum of participants the chance to communicate their findings. However, the usual time allotted, i.e. twelve minutes talk and three minutes discussion, does not allow in-depth interaction between speaker and audience. Depending on the material, it may be difficult to project more than ten slides, even if they are well prepared. The explanation of one extensive table may take up all of the speaker's time. Hand-drawing of overheads is usually not a good idea. In all, only experienced speakers may get their message across in twelve minutes, but they will rarely give Short Communications. The weakest point of these presentations is the discussion period. If the data are stimulating, it will be too short; and it takes only one long-winded talker to waste everybody else's time with a single comment.

To compensate for these problems, one can schedule a general discussion period after several presentations. Unfortunately, this is not always the solution either. Speakers may have disappeared, perhaps to catch a talk at a parallel session. Members of the audience may have forgotten what they wanted to ask (only to remember it hours later). Sometimes, the general discussion period may be too short because the preceding talks overran (despite explicit instructions to chairpersons and presenters). Last but not least, before lunch or evening breaks, some speakers have a pathetic urge to run away early, general discussion or not.

For the organizer, *parallel sessions* with Short Communications can turn into nightmares. If one session does not stick to the exact time frame, angry people will be rushing between overlapping lectures. To minimize this potential problem, Short Communications should be scheduled at least three minutes apart; and, if time is at a premium, according to the following pattern: (*a*) talk: twelve minutes; (*b*) discussion: five minutes; (*c*) intermission: three minutes. If this is not feasible timewise, they can be grouped into blocks of four that are separated by one or two breaks, lasting ten minutes or longer. These breaks are then 'buffer time.' Whenever possible, however, the time for the talk itself should be extended to 20 minutes. For more on Short Communications, see Section 9.2.5.

Let us now ask an almost heretical question: Why, in the age of Poster Sessions (see below), should there be Short Communications at all? As the latter tend to be unsatisfactory for a meaningful exchange of ideas, there isn't really much justification for their continuation. Perhaps, you may wish to retain them to help colleagues obtaining travel support which they will not get for poster presentations. This may not be a good reason, though. An organizer should be able to word an invitation so that the term 'presentation' appears in a non-committal way; which means that it can be interpreted to refer to a talk or poster presentation. On the other hand, there may be some sense in Short Communications at 'slave markets' where graduate students perform for potential buyers. Perhaps then, 'Graduate Student Sessions' should be retained by some societies. Also, at regional meetings with a benign audience, Short Communications can be the first chance for graduate students to speak in public.

3.1.1.5 After-Dinner Talks

After-Dinner Talks usually have an audience that is captured rather than captive. They can cause indigestion when people are pinned down at their tables in a postprandial state. Even worse, After-Dinner Talks may interrupt interesting conversations. To be sure, nothing is wrong with short, humorous speeches, thanks to the organizers, toasts to honored guests, etc., between courses or after dinner. However, a lengthy, idiosyncratic walk down memory lane can be a torture for everyone except the speaker.

If a dinner is held in honor of a distinguished person, the time for extended toasts and roasts is when drinks have lifted the spirits, and *before* food is served. It is not pleasant to listen in front of a glass with ice water and dirty dishes while the bus boys loom in the corners, ready to dash in the very moment the talk is over. It is definitely more gracious to phase out the event with conversations and beverages.

3.1.1.6 Closing Lectures

Closing Lectures are not necessarily a good idea. The speaker is expected to cater to the ego of all who gave talks during the meeting. This means he has to sit through virtually all sessions and take notes, and do miracles in the case of parallel sessions. It is not easy to prepare meaningful closing talks. They become an insult to the intelligence of colleagues when a speaker tries to cover up his ignorance with flattery. Some Closing Lectures make me shudder in retrospect: speakers relying on abstracts that were outdated at the time of the meeting, praising their friends explicitly, and referring to other presenters superficially without mentioning names. Unless a meeting is very focused, it is unlikely that, in this age of specialization, anyone can do justice to fifty investigators who presented data during a four-day conference.

In general, Closing Lectures may be more appropriate for short meetings, especially of the regional type. After several days of a major meeting, with a number of scientific highlights, people may be tired and skip the talk.

3.1.2 Poster Sessions

Poster Sessions have developed into one of the most effective means of scientific communication (see, e.g., R. A. Day *How to Write & Publish a Scientific Paper*, 4th edition; Oryx Press, and Cambridge University Press; 1994), and they should replace Short Communications as much as possible.

One major drawback with posters is the time it takes to prepare them. Also, it is more convenient to travel with slides and notes than pieces of cardboard. Recently, posters have appeared which consist of a single piece of plastic material that can be rolled up. Perhaps this will be the elegant way of the future.

The number and quality of posters and the facilities for Poster Sessions have increased substantially during recent years. Since organizers of scientific meetings often provide detailed instructions for the design, the preparation time for a poster should generally not exceed two to three days, especially when versatile computer programs (and/or technical support) are available.

Satisfactory posters can be created by following cookbook-type instructions (see Appendix A). However, outstanding posters result from a combination of concise language and a few well-designed and well-selected illustrations. This requires intellectual discipline, imagination, and some experience. To foster the continued development of such posters, scientific societies should encourage younger scientists with special prizes.

The best posters cannot make up for poor facilities. For the organizer, it is therefore imperative to visit the prospective site for a Poster Session and verify the following:

(1) Is there a sufficient number of poster boards? Do not trust the information given by the management of your meeting place. Go and count the number of boards yourself. A single missing poster board can become a big headache, especially when the author comes from far away.

(2) Is the quality of the poster boards adequate? Are they high enough, or do you have to get on your knees to read the texts? Are they made of appropriate material, strong enough to hold heavy pieces of cardboard without breaking; or are they made from some type of wood or artificial material that can hardly be penetrated by tacks or push pins? Note that there are excellent poster boards on the market; they are covered with burlap-like cloth, and provided with wheels.

(3) Are the poster boards of appropriate size? Are they big enough and are *all* of them of the same size?

(4) Is the room of adequate size? Insufficient space for poster presentations is a common problem. Make sure that the distance between two rows of posters will be at least 3.5 meters (about 3.5 yards). Remember that you cannot use the reverse of poster boards when they are placed against a wall. To be safe,

make a drawing of the room and place on it all posters with their numbers.

(5)   Is the arrangement of the poster boards adequate? The worst mistake in arranging posters is the creation of room-saving 'cubicles' where three posters cover three sides, leaving the fourth side open for access. Usually, this means that only two or perhaps three people can discuss a poster at a time, while the presenters and those interested in the two other posters must await their turn.

(6)   Is there enough illumination? If you are depending on daylight, is the sunny side too bright, or the shady side too dark? If the room has electric light, will all posters have sufficient light, or will there be underilluminated backrows once the poster boards are set up?

(7)   Is there space for refreshments? If poster sessions are followed by lectures, alcohol-free beverages may be preferable. Otherwise, a complete bar will not hurt. Remember that the costs for the bar may have to be included in the budget. In general, coffee and soft drinks should be charged to the budget of the meeting, while alcoholic beverages are paid for by participants at a cash bar.

Once you have ensured, by personal inspection, that the facilities for the Poster Sessions are adequate, pay attention to the following:

(1)   Avoid competition with Short Communications. Otherwise, people will present their best data during the talk. This is another good reason to eliminate Short Communications from your program.

(2)   Limit posters to a maximum of two per *senior author*. This cuts down on (*a*) presentations of inferior or less interesting material, (*b*) empty poster boards when authors do not show up, and (*c*) unmanned poster boards, because a person can only be at one place at a time. If a research group submits five posters and only one member comes to the meeting, up to four posters may be unmanned.

Sometimes, a participant wishes to display a poster whose senior author will not attend the meeting (a typical example would be a professor whose graduate student cannot obtain travel support). This should be permitted, provided that (*a*) the participant is a co-author, and (*b*) the poster is counted as if the presenter were the senior author.

(3)   Accept posters only if the abstract is submitted simultaneously with, or after, payment of the registration fees. This prevents empty poster boards caused by those who never intended to come to the meeting, but want to get their abstract published.

(4)   If there are many posters on closely related topics, schedule them in small groups (not exceeding ten, if possible) for different times. This will provide more time for discussions, and a better chance for the presenters to visit each other's posters.

(5)   Break the Poster Sessions down in two periods: (*a*) a preview time *without*

the authors; and (*b*) a poster discussion time *with* the authors present. Encourage the authors to have hand-outs (abstract and perhaps some additional information) available during the preview time.

Why a preview time? Recall that people hesitate to enter a shop when the owner stands in the door (I have heard of a place that went bankrupt for that reason); and that presenters of posters often scare people away by staring at them.

(6)   If the participants of your meeting have a wide range of interests, estimate the preview time according to the following suggestions (or any better method), and allow at least twice the time for the discussions with the authors. Allot 30 seconds average time per individual poster since only a small percentage of posters will be of special interest to a given participant. Assume that you have 100 posters covering 10 topics, and that a typical participant will inspect no more than about 10 posters in detail. On an average, an interested person will need about 5 minutes for a well-designed poster. When 50 minutes are available for preview, the participant will have 10 × 5 minutes for the posters. For practical purposes, you may assume that 45 minutes will suffice when the posters have been set up well before the preview time (for details, see Sections 5.3 and 17.3).

These times are based on personal observations during biomedical meetings. If the number of topics is reduced and the number of posters per topic increased, more time will be needed (even though the total number of posters will not change). The same holds true if posters are poorly prepared (e.g., too much text and/or large tables).

It is difficult to suggest specific times for poster sessions when a meeting deals with a specialized topic. However, a simple rule applies: in case of doubt, allow more time.

Recently, I learned of a remarkable way to gain more time for outstanding posters: several posters were selected for an in-depth discussion in a special session. One can safely assume that this splendid idea created a few very happy and many unhappy people.

(7)   Try to ensure the presence of the authors during the poster discussion time. For whatever reasons, some people will put up their posters and disappear. To prevent this, the announcement of a recent symposium stated that the authors have '*Anwesenheitspflicht*' ('mandatory attendance') during poster discussion time. I don't know if this Prussian-style order had an effect, but I can't come up with a better idea. Probably, the signing of a pledge would not do much good; since there is often enough space at the bottom of the abstract form, you might try it.

(8)   Mail concise instructions on the poster size: spell out length and height. If you just write '150 × 220 cm,' people may come prepared for the wrong dimensions. Spell out clearly: '150 cm width × 220 cm height'; however,

deduct at least 60 cm from the height if the board touches the floor. Give the dimensions *also* in inches.

(9)   Mail concise instructions on the layout and mounting of posters (see Appendix A).

(10) Announce in the program the exact times when posters must be mounted and removed. If possible, have posters put up before the morning session so that people can inspect them during breaks, or when they are not listening to talks. This will cut down on 'traffic jams' during the scheduled preview time.

3.1.3 Colloquia and Round Table Conferences

Colloquia can be organized in different ways. When there is a large number of participants, 'Workshops' are appropriate, but if there are fewer participants, it is possible to organize Round Table conferences. Perhaps, the most distinguishing feature between different types of colloquia is the presence of an audience. For simplicity, let's define Colloquia *without* a general audience as Workshops, and consider them in the next section.

At a Colloquium as defined here, the audience will participate from a certain point on, or at least will have a chance to ask questions. Occasionally, though, high priests of science do not wish interference by common man. Hence, you had better announce in advance that all Colloquia at your meeting will involve the audience to some degree.

A Colloquium is usually doomed to failure if there are no time limits on the individual presentations (especially when these include slide projections); or if the moderator capitulates before the unsuppressable verbiage of fellows who confuse a question with a sermon.

A *sine qua non* for a successful Colloquium is a challenging topic: the more controversial or confusing the issue, the better. Of equal importance is a *good* moderator. If none can be found, you should abandon the idea of a Colloquium. The selection of participants is not to be taken lightly, either. Undisciplined marathon talkers and inflexible doctrinarians must be excluded; qualified participants with differing views, or at least different approaches, have to be identified. The maximum number of people on the platform, including the moderator, should be five if the moderator wishes to talk on his own work; and six if he prefers to give only a short introduction to the Colloquium. Note that this number has been used for the example in Section 17.3, and also for the estimated time (100 minutes) given in the 'General Schedule' (Appendix B).

Furthermore, *clear rules* have to be established. The introductory talks of the panelists should not exceed ten minutes, and shorter remarks should be encouraged. Recall: 6 × 10 minutes = 1 hour! Even though the total time available may not exceed two hours, sufficient times (at least 30 minutes) for the general discussion with the audience should be included (see Appendixes C and D).

The question of visual aids must be addressed: Should slide projections be permitted, even though this invites attempts to go overtime? My advice would be: If they cannot be avoided, allow only two slides per panelist. On the other hand, provide a blackboard or flipchart since (*a*) these can help to make a point, and (*b*) may stimulate interactions. An overhead projector is better avoided since people tend to overuse it. Time permitting, each platform presentation should be followed by a five-minute (or longer) discussion, during which the panelists have the prerogative to ask questions; however, questions from the audience may be admitted if the five minutes are not used up by the panelists. Such brief discussions keep the interest of the audience going.

Perhaps, the ideal Colloquium would be as follows. Five minutes introduction by the moderator; five 10-minute presentations of platform panelists, each followed by five minutes discussion (= 75 minutes); and 40 minutes of general discussion. Total time: two hours.

A Colloquium will greatly benefit if the moderator identifies important questions as they emerge and then leads the general discussion to a point where he can demand *clear answers*. For example, he may ask: 'Today, we have heard so much about the effects of hormone X on the prenatal development; do we have any proof that it actually has a physiological role before birth?' If the answer is a unanimous, surprised 'no,' the colloquium should be considered a success just for identifying this one, critical issue.

3.1.4 Workshops

Workshops flourish in a humorous, congenial atmosphere. Since dullards can turn any event into a wake, selection of the right leaders is important. As science becomes more and more specialized, the need for personal interactions increases, and the value of workshops is growing proportionally. Workshops deserve a much larger share of scientific programs than they have received traditionally.

3.1.4.1 'Hands-On' Workshops

Workshops with hands-on experience (e.g., 'New instruments for the clinical laboratory,' or 'Simplified quality control of urban water supply') are not considered here in detail. They are mostly events of their own which, by necessity, vary greatly and require much custom-tailoring. When connected with another meeting, Hands-On Workshops are best scheduled directly following the main event. It is unfortunate that they are sometimes held in parallel with sessions; and worse, even at locations distant from the congress center. This creates a dilemma for people interested in attending sessions *and* the workshop. After all, participants are entitled to get their money's worth if they have paid fees for both the main meeting and the Hands-On Workshop.

3.1.4.2 Discussion Workshops

This type of workshop should be by invitation only. Once the topic and workshop leader(s) have been identified, up to 25 participants can be selected. It may be preferable to restrict the number to 20 since the less participants, the more time for each one to say something. If Discussion Workshops are held in connection with a meeting, cooperation between the organizer of the meeting and the workshop leader(s) is critical: participation in workshops may have to be balanced against, and coordinated with, involvement in other events, such as lectures, Colloquia and other workshops. This can become a formidable task, considering all the egos that are so easily bruised. Discussion Workshops differ from Colloquia in that there are no privileged panelists. Their purpose is manifold: in-depth discussion of scientific problems; establishment or deepening of personal contacts; exchange of experience; creation of joint projects. This means that the participants must be given a maximum of time for interactions.

As in the case of Colloquia, slide presentation should be kept to a minimum, or excluded, and use of blackboard and/or flipcharts encouraged. If an overhead projector is permitted, strict time limits must be set.

Structured Workshops involve short introduction to topics that are followed by discussions of flexible duration, with a maximum time limit. The introduction should briefly outline questions related to a specific topic, but never turn into a talk. If the discussion of a topic terminates before the allotted time, the next presentation can follow instantly. As a rule of thumb, one can probably assign 30 minutes per topic (5–10 minutes introduction plus 20 minutes discussion). After a number of topics, a general discussion period (the longer, the better) is recommended since participants may wish to bring up issues that have not been covered previously. It is also advisable to schedule breaks (lasting 5–10 minutes) between the topics so that a discussion of interest to only a few persons can continue for a little while.

Non-Structured Workshops serve to discuss data, exchange experiences and ask questions. This informal variant is sometimes more fruitful than a structured workshop since it makes it easier to obtain specific information. All it takes is a skillful leader who ensures that all participants have a chance to bring up matters they wish to discuss. However, nobody is required to make a presentation. In order to maintain informality, again only a blackboard or flipcharts should be used. Perhaps, the optimal number of participants is about twelve, with a minimum time of two hours.

3.1.4.3 Socratic Workshops

There is yet another type of workshop, and I highly recommend it. It allows fruitful, informal discussions among a larger number of participants, and encourages younger scientists to speak up. Furthermore, it is the perfect justification for formal letters of invitation.

The Socratic Workshop has its roots in the ancient symposia, and we introduced it under the pseudonym 'Evening Workshop.' 'Symposium' is of Greek origin, and it means 'drink-together.' This is exactly what many Greek and Roman scholars did after a good dinner: lying on comfortable longchairs, engaged in animated discussions (and frequently other fun as well). Most notable among these wise men were the disciples of Socrates. Their publications are still quoted, even read, after 2400 years; which is unlikely for the legacy of most Nobel Laureates. Interestingly, Socrates himself never wrote a paper.

The idea of the Socratic Workshop is good science in a relaxed atmosphere of wining and dining (see Appendixes E and F). After a full day of mostly structured presentations, body and mind deserve something better than another semi-Spartan event.

We tested the Socratic Workshop at an international symposium. There were a total of twenty-two of these 'Evening Workshops,' and the number of participants ranged from five to about thirty. On four evenings, up to seven workshops were held in parallel in selected local restaurants, away from the congress hotel for a change of scenery.

If a meeting catalyzes an unprecedented number of new personal contacts and cooperations, and the participants return home with new ideas, feeling intellectually rejuvenated (as one colleague put it), then it must have been worth attending. We achieved this goal, and by general consensus the Socratic Workshops were instrumental in the success.

The recipe for Socratic Workshops is as follows:

(1) Identify appropriate topics. Consult with your organizing committee, and contact informally (preferably by telephone) experts, if your committee is not familiar with a field of research.

Consider, *inter alia*, the following: (*a*) upcoming new fields of research; (*b*) active fields where a glut of data requires critical evaluation; (*c*) neglected fields that need a boost; (*d*) two fields that have grown apart but could mutually benefit from an integration of the existing knowledge; (*e*) areas of research where personal contacts are missing; (*f*) more narrow areas of research where old acquaintances will appreciate the chance to exchange their latest ideas (but make sure that it is not only 'old boys' at the exclusion of other, especially younger, scientists).

(2) Estimate the probable number of persons interested in a given topic. Try to assign no more than twenty participants plus two leaders to a workshop, keeping in mind that there may be ultimately more rather than less participants. If necessary, split a topic into two more narrow ones and schedule them for different evenings. This will enable people with a wider interest to attend both workshops, and allow others to focus on more specific questions during one of the evenings.

(3) Compile a list of workshops and try to fit them in the available time slots. This can become mind-boggling since you must avoid overlap of workshops on related topics, and the more workshops that are scheduled, the more likely that people will be interested in two that are held on the same evening. Your job will be easier if you can spread the workshops over more evenings.

(4) Identify prospective meeting sites (restaurants or catered private places) for the workshops, and make sure that all of the following criteria are met: (*a*) reasonable prices; (*b*) sufficient and undisturbed space for the workshop; (*c*) adequate selection of good food (especially if you anticipate participants that may not eat pork, beef, or any meat at all); (*d*) clean toilets; (*e*) friendly and efficient service; (*f*) good location.

Sneak a look at the kitchen, and check how much time they need to serve a dinner on a crowded evening. Talk to the owner (or person in charge) about the purpose of your visit only when you have paid your bill. If he is really interested in catering your workshop(s), agree with him a later date for the discussion of details.

(5) Assign the workshops to the selected places. If workshops on similar topics are held on different evenings, schedule them for different restaurants; this increases the chances that people will be exposed to more variety. Assign ichthyologists to a seafood place and entomologists to a garden restaurant with bee hives; the familiar settings should make them comfortable. For psychologists and some clinicians, though, this strategy may not be attractive.

If you want to make it easy on yourself, you can skip the use of questionnaires as outlined in the following, and concentrate only on the points dealing with workshop leaders and restaurants. However, if you want to stimulate an optimum of interactions, proceed as follows:

(*a*) Design an example of a 'statement of research interests' (see Appendix G) that will be mailed to all workshop participants. The participants will send corresponding 'statements' with their own interests to the workshop leaders; in turn, the leaders will mail sets of the compiled statements to all participants of their respective workshops.

The purpose of the 'statements' is to inform the participants about each other's research and technical expertise, need for information on specific items, and interest in collaboration.

(*b*) Identify two leaders for each workshop early, if possible three years before the meeting. You will need people who are willing to (*a*) try something unconventional, and (*b*) spend time on the preparations. If someone tells you that he must think about the assignment, gently withdraw your offer. It will be bad hunting if you must carry the hound to the woods. At international meetings, one would like to have one older and one younger person, of different sex and nationality, as workshop leaders.

(*c*) Familiarize the workshop leaders with their chores, and make sure that they work together. Give them your list of prospective participants (if you have any), and ask them to add whoever they wish as long as the total number, hopefully, does not exceed twenty. However, emphasize that nobody must be blackballed; and that, on the other hand, the invitation to a workshop does not entail financial support (unless, of course, you can provide it). Finally, arrange the mode of payment for their mailing expenses. Make sure that they have sufficient funds, and that the mail will be sent by airmail, whenever necessary.

(*d*) Well before the meeting, confirm that your workshop leaders are willing to cooperate. If someone tries *now* to make the acceptance of the job contingent upon special privileges (e.g., unreasonable financial support), replace him or her right away.

(*e*) Immediately following this confirmation, mail invitations, information and requests for 'statements of research interests' to all prospective workshop participants and inform the workshop leaders.

In a cover note to the participants (see Appendix E), point out that (*a*) the 'statements' must be returned to the leaders of their respective workshops by a deadline (four months before the meeting seems reasonable); (*b*) copies of a completed set of questionnaires will be mailed by the workshop leaders so that all participants have the information on their colleagues *before* the workshop; (*c*) this information will allow interactions to begin at the very moment people arrive at the site of the workshop.

(*f*) Six weeks before the meeting, ask the workshop leaders if the compiled questionnaires have been distributed to their participants. If this has not happened, impress upon them that the participants must be well informed about each other before the workshop starts. If the worst comes to the worst, the sets can be handed out at the registration desk.

(*g*) Also six weeks before the meeting (and just before the program goes to press), confirm with the selected restaurants and/or caterers the precise dates and times of the workshops, and inform them of the probable number of participants. This is also the moment to check that there will be no surprises with the restaurants, such as vacations, change of ownership, ongoing renovations, permanent closure, etc.

(*h*) One day before the meeting, have flipcharts, easels and markers in different colors delivered to the rooms where the workshops will be held. Do not allow slide projections or overheads unless the subject matter makes them indispensable. For workshops with a few participants, have a round table set up. For a larger group, a horseshoe-type arrangement may be best. This is an opportunity for the workshop leaders to come along and familiarize themselves with the site.

(*i*) Recommend the following general schedule for the evening: (1) 30 minutes for drinks; (2) brief self-introductions followed by 90 minutes or less (seated) dinner time, depending on the speed of service; (3) 60 minutes or more general discussion (seated) with beverages readily available. Total workshop time: at least three hours.

The workshop leaders should have flexibility in running the event, keeping in mind that conversations are the most important purpose of the evening. Everybody who wishes to raise an issue for general comment should be given a chance to do so; the best way to ensure this is to list the proposed topics on the flipchart at the beginning of the evening. This way, everybody knows the agenda, and the leaders can bring up point after point during the general discussion, whenever the moment appears right. There is no need to close the evening officially. Rather, people should have a chance to enjoy each other's company as long as the restaurant remains open.

One word of caution: the organizer and the workshop leaders must insist that participants consuming alcoholic beverages will not drive a vehicle after the workshops. Unless their accommodation is within safe walking distance, the participants should return by taxi, or bus if one has been hired for the evening.

3.1.4.4 Open Workshops
Perhaps, a better term would be 'General' or 'Open' Discussion. However, since this type of discussion is often referred to as a workshop, let's use the above term. In essence, Open Workshops discuss one or more topics that have been announced in the program. They are often held with a large audience in a lecture room; sometimes, the audience is restricted to those registering beforehand, but generally, there should be enough seats to accommodate all persons interested. The success of such workshops depends on the skills of the moderator, and luck. The best moderator may run out of luck when some people try over and over again to monopolize the discussion, or to deliver lengthy monologues. If these problems have to be anticipated, it may help to limit the time for individual questions and answers, using some kind of alarm device.

3.1.5 Automated slide presentations, videotapes and films

Automated slide presentations can be used for a wide range of topics, from the introduction to a National Park to the demonstration of a surgical procedure. Depending on circumstances and equipment, they require a screen (photographic slides), or television monitors (e.g., laser disks), and usually a sound system. Overall, automated slide presentations are not too common at scientific meetings, and they are more frequently used for short commercial sales pitches. However, they can be very effective when complex non-moving structures, such as geological maps or histological preparations, are discussed in detail. When given via good television monitors, they are ideal for some types of Hands-On Workshops since they do not require darkening of the room. A typical example would be a session during which participants compare preparations on their microscopes with items that are discussed on monitors.

Movies (films or videotapes) may be essential when it comes to the analysis of dynamic processes, from ovulation and muscle contraction to the gait of a horse and the merging of galaxies. If movies have to be backed up by still copies of certain frames, two screens or monitors will be needed.

Whenever automated slide presentations or movies are used, the precise nature of the necessary audiovisual equipment must be known as early as possible since some rooms may not be suitable for all types of projection. Hence, information on the available facilities, and a questionnaire concerning the necessary equipment, should be mailed with the invitations.

Scheduling of movies can be a challenge. When they are shown as films in dark rooms, it may be difficult to keep the attention of an audience for several consecutive presentations, especially when a room is poorly ventilated. It may help to alternate films and videos, since the latter are usually shown without dimmed lights, and certainly 'lights-on' time with discussions and/or breaks between movies should be scheduled. However, everybody will be better off if videotapes are used.

Alternatively, one can intersperse movies between lectures so that the audience is exposed to more variety. In general, however, movies should be scheduled towards the end of a session. If technical problems occur, it may be difficult to shorten an automated presentation in a meaningful way. A further consideration is that videotapes tend to give less problems than films; an older film definitely belongs in the last time slot of a session.

Entertaining movies or slide presentations are best scheduled for times when the audience is less able to concentrate. This is usually in the late afternoon; or after dinner, if evening sessions are held.

For societies with members of wide-ranging, in particular faunistic, botanical, geographic or astronomical, interests, special rules apply. Their meetings are sometimes attended by many amateurs and accompanists who do not go through the all-day stress of scientific sessions. For the benefit of these participants, travelogues and entertaining movie presentations should be arranged for some of the earlier evenings, after dinner. These events should not last too long so that people have afterwards sufficient time to socialize.

3.1.6 Scientific and technological demonstrations

Depending on the type of meeting, the organizer may have to anticipate requests for special scientific or technological demonstrations. For example, participants may wish to set up models of molecules, crystals or organ systems; combine posters with slide presentations; or display new instruments. These demonstrations can enliven Poster Sessions, and they should be encouraged if space is available. Obviously, they must be kept in mind when meeting facilities are selected.

If space is limited and there are many requests for such demonstrations, it may be necessary to schedule most or all of them simultaneously as a special event. In this case, the last day of the meeting may be the best time since, if necessary, the poster boards can be removed on the preceding evening to make space.

3.2 General and Business Sessions

The agenda of these sessions should be mailed in advance. A lot of time is lost when people are handed important information at the beginning of the session, or if they are poorly informed on matters to be discussed. Depending on the occasion, tradition and/or society, General or Business Sessions vary greatly. Nevertheless, as a general rule, routine and simple matters are scheduled for the earlier part of the sessions, and new or complex issues for later. The total time will depend on the agenda and the topics; however, the time for discussion is often insufficient. Hence, a manipulator will sandwich a Business Session between two other events if he wishes to keep it short.

Meetings in the USA may be run according to 'Robert's Rules of Order Revised for Deliberative Assemblies.' However, since these rules are not used in many countries, international societies should run their business sessions by the rules of common sense.

What are Robert's Rules anyway? According to the Encyclopedia Britannica, Henry Martyn Robert (1837–1923) was a US army general. As a young lieutenant, he became upset during a church meeting in which there was much controversy. This experience led to his pocket manual which was first published in 1876. Today, the revised edition of the year 1915 is the authoritative guide. It consists of two parts: 'I. Rules of Order. A compendium of parliamentary law, based upon the rules and practices of congress'; and 'II. Organization and Conduct of Business. A simple explanation of the methods of organizing and conducting the business of societies, conventions, and other deliberative assemblies.' The rules are neatly outlined in 75 paragraphs, the last one dealing with 'Trial of Members of Societies.' No wonder the fellow was made a general.

Depending on the customs or rules of an assembly, it may be necessary to take minutes of the business sessions, and to have them approved at the next meeting. If so, it will be helpful (and good for documentation) to tape the sessions even if a secretary takes notes. Which means that the organizer must provide a tape recorder and sufficient tapes.

A Business Session that ends without a decision on an urgent issue can be a big headache for an organizer (unless, of course, he planned it that way). Frequently, the problem is that people say nothing in many words while time is running out. Democracy is fine when it works, but it can be self-defeating. Hence, whoever moderates the business meeting may be forced to resort to 'Plan B'; i.e., recruit in advance reasonable colleagues who will help to reach a decision. Of course, when this strategy becomes obvious, he may be accused of Machiavellian tactics. That could be a small price to pay considering the alternatives such as: the disintegration of a society; financial ruin of a journal; or, at the least, months of expensive correspondence via international air mail, paid by membership fees.

If all fails in a Business Session, the moderator may yet be able to have a committee approved that has the authority to make the decision. But then again, the committee may resolve nothing in a quagmire of arguments; or, if not set up properly, end its deliberations in a stalemate (e.g., when it consists of an even number of members without a tie breaker). For more on committees, see Chapter 10.

From the above deliberations, it would appear imperative to prepare a Business Session as carefully as possible.

3.3 Forums

Forums deal with issues of interest to all participants, and perhaps also to the general public. They can cover almost any topic, ranging from the local to the international level, and from the philosophical to the very concrete. Examples are: a radioactive waste site; academic freedom; funding for research; changes in educational programs; interdisciplinary research programs; international cooperations; abortion and population control; animal rights; pollution; habitat destruction. If the issue is serious, a Forum may end with a resolution which all present will vote on; and/or a document signed by representatives, or all in favour of its contents.

Usually, a Forum begins with a few introductory remarks by the moderator, followed by one or a few short presentations by experts on the topics to be discussed. If the issues are controversial, the moderator must remind the audience at the outset that the presenters cannot be held responsible for the facts they report; and consequently, should not be treated with hostility. Of course, if an independent thinker, a hero or a masochist agrees to defend an unpopular decision, it may take more than a friendly reminder to keep the situation under control. Whenever possible, the moderator should run a Forum so that it will close on a conciliatory note (see also Section 9.2.4).

3.4 Commercial exhibits

As you prepare a smaller meeting, one or a few publishers may ask permission to display books. This should not create a problem if: (*a*) space is available; (*b*) no more than a few tables are needed; and (*c*) you have full authority over the meeting area. You may either allow them to set up a little display on an informal basis (if they have someone who will be there), or you may suggest that they do it through one of the local bookstores. And if that is your style, you may even get sponsorship from them. However, do not take any responsibility for the books; otherwise, you may have problems when one of them disappears. Corresponding rules apply for minor demonstrations of equipment.

If you do not have full authority over the meeting area, it may create a grey area of rights and responsibilities. This is one of the reasons why a contract with the owner of your meeting place is recommended (see Section 6.1). Matters become very complex when a meeting involves major exhibits that include stalls for books, instruments and merchandise, not to mention live animals.

Ideally, you should have a contract which explicitly allows you to rent out exhibition space if you choose to do so (even if you got all meeting facilities for free). If this issue comes up in negotiations, one of your bargaining chips should be free coffee in the exhibition area which will be paid for out of your meeting budget. This way, the owner of the exhibition area makes directly, or via his caterer, money from you; while you, in theory, make money from the exhibition. The final outcome of this deal may depend on the price of the coffee. If you don't watch it, the charges for the coffee could be so outrageous that you will be the ultimate loser.

Handling of a major exhibition (with many thousands of dollars in rental fees) requires professional experience. You may have to provide decorated booths, and for this job you need an exhibit decorator. In a US city, the exhibit decorator may be picked locally after a bid has been sent out. In turn, the decorator may recommend an 'exposition service' which handles the shipment of the exhibit items. Appendix H summarizes important points to consider for a contract with an 'exposition service.'

Of course, once you charge money for the stalls, people demand things in return. A friend of mine with ample experience assures me that this often leads to nasty arguments. Also, you may have to provide guards who must be paid from your budget. So, it is not just the free coffee that must enter your balance sheet.

To sum it up: major exhibitions demand a lot of time, and legal, logistic, financial and other considerations. If you don't have pertinent experience, you definitely need guidance. If possible, consult with the person who was in charge during the previous meeting of your type. You will be fortunate if you can rely on a society's executive officer who handles exhibitions routinely.

3.5 Disruptive demonstrations

In free societies, everybody can speak up. Unfortunately, this can happen during meetings when outside groups want to be heard. To avoid serious confrontations, precautions must be taken when disruptions are expected. I have experienced two potentially explosive situations in which reasonable compromises were found. They may give some guidance. Of course, police protection and other measures may be necessary when physical violence or destruction of property is likely.

The first incident occurred at a scientific meeting in the USA during the Vietnam war. The organizers had been forewarned, and thus were able to instruct session chairpersons and speakers. In short, we were told to stop our sessions and listen to ten-minute speeches which would deal with the insanity of the war. When the students showed up during my lecture, I stopped instantly, and the session chairperson asked them to make their point. We listened, they thanked the audience for their attention, and we continued the session.

The other incident happened at an international meeting in Europe. Animal rights activists had planned a major confrontation, and they greeted the participants on their way to the opening session with shouts like 'murderers.' Subsequently, however, a compromise was reached, and the activists were permitted to set up a monitor, showing a continuously running display of cruelties to animals (the original tapes had been stolen from a research laboratory). This demonstration was in a hall outside the area which was restricted to registered participants. As far as I am aware, there were no further confrontations at this meeting.

4

Social events: something for every taste and budget

4.1 Scheduling and options

4.1.1 Minimeetings

Next to the quality of the presentations, the social events will decide the success of a scientific meeting. If a meeting lasts only for one half or a full day, extended coffee and/or lunch breaks should provide an opportunity for informal mixing of the participants. This is particularly important in a city where people flee to their suburban homes at 5 pm. Nevertheless, you may find some time for postsymposium drinks. Under more relaxed conditions, you may be able to arrange a seated dinner, or a catered buffet for some of the local participants and out-of-town speakers who are staying overnight.

4.1.2 Meetings of longer duration

If a meeting with out-of-town participants extends into a second day, the evening of the first day should be reserved for a social event. In most cases, an all-evening party with a buffet and a selection of beverages will be perfect. If costs are a problem, tickets for food and/or drinks can be sold.

For meetings lasting three or more days, an informal reception or welcome party on the first evening and a major social event on a later evening are generally expected. There are two schools of thought concerning the timing of the major event. Some people prefer to have it on the last evening; others would like to have it earlier because they intend to leave during the afternoon or evening of the last day. For the benefit of the tired organizer and his staff, and of all participants who hate morning sessions after a long festivity, the major event would best be scheduled for the last day. However, the organizer may have to bow to the tradition of a society and schedule it one day earlier. This may also be mandatory when the price of overnight accommodation forces participants to cut their stay as short as possible, or when no events are scheduled for the afternoon of the last day.

For meetings with four full days of sessions, more social events are called for. As a rule, the afternoon of the third day should be used for an excursion; unless, of course, you can extend the meeting to last five days, and use all of the third day for the excursion. If the meeting has five or six full days of sessions, the obvious choice for the excursion would be the fourth day, so that three days of work are followed by a day of relaxation, which in turn is followed by two or three more days of work.

A special situation arises with meetings that are attended by several hundred, or even thousands, of participants. Frequently, the large number of people will make a simultaneous social event for all impossible. This may not always be bad since large conventions are often nothing but an umbrella for different meetings whose participants have already a hard time to find old friends or colleagues. At many major meetings, therefore, it may be best to have the individual divisions or sub-groups decide on their own social events. However, the overall schedule must provide the necessary time slots; the events must be coordinated so that, e.g., sufficient dining rooms are available; and probably, the person(s) in charge of the local arrangements will also have to assist with details, from the selection of the menu to transportation.

Occasionally, the organizer of a large meeting may still decide to reserve a half or full day for excursion(s). In this case, early and careful planning will be necessary. Even in a large city, it may be difficult to hire sufficient buses, and there could be serious logistic problems at the destination(s). How many people can you pack into a museum or castle?

Sometimes, however, there may be a simple solution. If, for instance, a major meeting is held in a city that is connected by railway with an interesting place, it may be possible to rent a special train; and for lunch and/or dinner at the destination, you may be able to reserve a sufficient number of restaurants. Also, an experienced tour operator will break up a large crowd into smaller groups that have their meals at different times.

Another type of excursion for a major meeting would be a trip on boats that are used to handling large numbers of tourists. How about a day on the Rhine, Danube, Hudson, Mississippi, one of the Great Lakes of North America or the Inland Sea of Japan? For further suggestions, see below.

For a meeting with about one thousand participants, or perhaps even more, an outdoor barbeque or cafeteria-style dinner may be possible, provided an *experienced* caterer can be found. Do not take chances with an outfit that will use your participants to learn the trade. Humans have a surprisingly long memory when it comes to bad food.

4.2 Suggestions

4.2.1 The setting

Depending on the circumstances, participants of a meeting may arrive throughout the day and evening preceding the official starting date. This is common at international meetings when airline connections vary. In this case, you may feel obliged to provide entertainment for the people who arrive early; and at the same time, you may be reluctant to spend funds on an event that cannot be attended by all participants who paid their full fees. The answer to your prayers may be a local sponsor for the evening, or at least free beer from a local brewery. If you have an official congress hotel, try to tap their management (don't hesitate: they'll get the money back in their own ways). Otherwise, you can run the evening as an informal 'no host' event with drinks paid at the bar.

Whatever the social events of your meeting, provide as many different settings as possible. Most people enjoy both beauty and variety, or at least originality. For practical purposes, parties preceding a meeting are best held close to the overnight accommodation. If you have the initial event around a swimming pool or near the ocean, hold the official reception on the following evening at a different place: a museum; a zoo; a castle; a historical town hall or manor; a park near a river or lake (without mosquitoes, please); an hacienda. If the major event of your meeting (probably a banquet) is held at the congress hotel, have cocktails in a place different from the ones where you had the 'no host' or welcome parties. Ideal places would be a courtyard or a veranda outside the banquet room. And if you plan to have Socratic Workshops, follow the above advice (Section 3.1.4.3) and look for interesting restaurants away from the meeting site.

Dinners and banquets are the perfect opportunity to insult people unintentionally. If there is a head table for certain people, others will feel they should sit there, too. You can even insult a whole nation if you exclude its supposedly most distinguished representative from that special table. On one occasion, the self-styled delegation leader from a European country left us in the middle of a conversation when the High Priests began settling down at the head table. A little while later, he returned with a burning face; poor baby, they had no place for him so high up. And as usual, you can top an insult: for example, when only the head table gets unlimited wine, while the lower ranks have to do with one bottle per table.

To put it in plain language: don't have a head table. Just reserve one or more regular tables so that you can seat your honored guests with friends and/or representatives of your society. If possible, seat people close to the microphone when they are supposed to give a talk or make announcements.

An invitation for cocktails or dinner in a private home is always appreciated. However, it has to be handled so that nobody feels excluded. At small meetings, this can be achieved if several receptions or dinners are held in parallel. At larger meetings, you may select people who can be identified with something specific, such as their research, or a home country. If the invitation is for a seated dinner, the host needs to seat the right people together. For example, it helps when quick-witted persons from Romance countries are interspersed with members of more reserved nations. Then, even Americans and Japanese may join in the discussion of controversial topics, something Europeans often savor to the fullest without taking offence.

At international conventions, the local consul of a foreign nation, or a member of his staff, may invite his countrymen to a reception. Or, perhaps the governor of a state or province may invite all foreign participants. If this must be anticipated, it has to be coordinated with the other events of the meeting. A tactful inquiry in time may prevent confusion later when the schedule has been finalized.

4.2.2 Food

Something with a local touch is often appreciated (exceptions granted). Try to combine it with a special setting. Here are some not too expensive examples: a salmon barbeque at a Canadian river; a clambake, or a luau, on a clean ocean beach; buffalo steaks on the prairie; Cajun-style seafood in Louisiana; a barbeque on the pampas; roasted pig, Breughel-style, in a Belgian barn; roasted lamb on a mountain meadow in the Balkans; a buffet with local meats and sausages, in a castle overlooking the Rhine river; casse-croute gaulois in a cave-turned-wine-cellar near Paris; buffet during a boat trip on a scenic river, lake or fjord; an authentic Chinese dinner in Hong Kong, Singapore or New York; home-made pies under the Southern sky, somewhere in Australia.

However, don't go overboard and consider 'prairie oyster' (bull testes) in Kansas; squirrels in Texas; haggis in Scotland; whale tongue in Japan; froglegs, kangaroo meat and songbirds anywhere. Some people may either dislike these specialities, or just take exception to them. And always remember participants who don't eat meat; make sure that special dishes will be available for them.

4.2.3 Beverages

Here also, local specialities are usually welcome. Of course, there must also be soft drinks for people who prefer something non-alcoholic. Beware of beverages that can harm those who are not used to them. Don't offer young, fermenting wine which tends to be a laxative; and don't offer wines that are sweet because either sugar has been added, or their fermentation has been stopped. They are for tourists who do not know what they drink – that is, until they get up from their chairs. There are many killers for the uninitiated: in particular, mixtures with rum, gin, cognac ('brandy') or bourbon; they return to their port of entry when they find too much company in the stomach. Strong cocktails may create problems, especially when meetings are held in hot weather, warm climates or at high altitudes.

On the other hand, a plain Spanish-style sangria is an excellent punch for receptions on warm evenings, if prepared with red wine (in North America, use California Burgundy), triple sec, orange juice and a lot of ice; however, you can add fruits and other liqueurs and increase proportionally the chance of problems. Also recommendable for summer evenings is a punch from the Rhineland, called there 'Bowle.' It is made from dry Rhine whine (or California Chablis) and simple, clean champagne, with fruit (or fresh extract of lemon peel) and moderate amounts of ice. Another punch for outdoor events is Lynchburg Lemonade, made from Bourbon, triple sec, lemonade concentrate (or lime and sugar), 'Seven-Up', and ice. Sangria, 'Bowle' and Lynchburg Lemonade have a great advantage: they can be stretched with ice when the supply runs low. For winter evenings, a hot Swedish wine punch ('*glögg*') or a '*Feuerzangenbowle*' (as served in Central Europe) can add a special touch to receptions. '*Feuerzangenbowle*' is better reserved for smaller groups since the preparation involves a burning sugarloaf, soaked with strong rum, that slowly drips into a hot mixture of red wine and citrus juices. Of course, some people prefer beer to anything else. So, just make sure that they can have it, too.

If you plan an international social event, remember that different cultures vary in their consumption of alcoholic beverages. Beer should always be available to those who prefer it to wine or hard liquor, even when meetings are held in countries which are major wine producers.

4.2.4 Entertainment

Music enlivens social events, as long as it is not depressing or boring, and does not interfere with conversations.

At cocktail parties and during lunch or dinner, cheerful local music can create a congenial atmosphere. A single accordion, bagpipe, balalaika, banjo, cymbal, guitar, mandolin, piano, or zither may be all that is needed to set people in a good mood. Budget permitting, a small band may be preferable if you have a large crowd. How about a Spanish '*tuna*' (band of students in historical costumes), Mexican mariachi players, a Hungarian gipsy band or Greek folk musicians? But beware, there are people who just cannot stand whining folk songs, especially when they drag on; not to mention oversentimental '*Kaffeehaus*' kitsch from central Europe.

Dance performances are in order, as long as they are of the right kind and well timed. Andalusian dances, dances of the cossacks, Bavarian '*Schuhplattler*', New Orleans-style tap dance or authentic Geisha entertainment (in Kyoto) are definitely a better choice than revived peasant dances from northern Europe, or belly dancing by amateurs that couldn't show you the Near East on a map. There is nothing wrong with a well-timed, not too lengthy demonstration of local folklore; or something close to it like a Mardi Gras show in New Orleans, or a Mummers performance in Philadelphia. During a banquet, the best moment for such shows may be towards the end, between courses, or immediately following the last course.

After a dinner or banquet, many people like to dance. It is therefore critical to pick the right band. An evening becomes a disappointment when musicians confuse scientists with hearing-impaired teenagers who spend hours limb-shaking; or, when they specialize in schmaltzy 'ballroom music' from yesteryear. Make sure that the band can play at a noise level that makes conversation possible, and that they have a *varied* repertoire that suits those who enjoy real dancing (with emphasis on livelier music).

At an outdoor barbeque, a little lesson in North American or Australian square dance, Mexican hat dance, Argentinian tango, Caribbean limbo or Polynesian hula hula can be fun; provided, however, people are not pressured to participate when they just like to watch. The same goes for sing-alongs with rocking in a Bavarian beer tent, or in a boat on the Rhine.

There is nothing wrong with a demonstration that does not hurt man or beast. However, in this day and time of 'heightened awareness,' it is easy to get into trouble. For instance, I would be reluctant to invite a falconer to demonstrate a bird of prey catching a rabbit or grouse, even at an ornithologists' meeting. On the other hand, trained cormorants catching fish by torch light (e.g., on the Nagara river in Japan) may be generally acceptable. I would avoid anything close to a bullfight, even if it is of the bloodless Portuguese or Provencal variety. 'Calf roping' contests should be less controversial, especially when sober cows outsmart soused cowboys or gauchos. Also, a performance with Andalusian or Lipizzaner horses, or by a Hungarian csikos (riding herdsman) should cause no protests (perhaps, not yet). A dinner at the local race course should also be generally acceptable. You may be even safer if you invite woodcutters who push each other from floating logs; one day, though, this may be construed as an endorsement of unnecessary paper production, and the same may go for Scotsmen throwing big logs through the air ('tossing the caber'). In short, you should consider yourself lucky if you can enrich your program with entertainment that does not ruffle somebody's feathers.

For many people, there is nothing like a good concert in the right ambience, be it a baroque residence or a well-designed modern music hall. Also, many participants of your meeting may enjoy a good stage performance, be it musical or otherwise.

During the warm season, perhaps one evening can be reserved for an outdoor concert. If the meeting is held during a cold or rainy season, an evening with a cultural event may be especially welcome. In the latter case, see if you can arrange a champagne reception (with some snacks) during a break, and/or a nice supper afterwards.

4.2.5 Excursions

Excursions may be professional, recreational, or both. By necessity, professional excursions will vary with the scientific discipline; in general, though, it is appreciated if they are phased out with a pleasant social event.

This is usually arranged for field trips. For example, excursions may terminate at a rustic restaurant serving local beer, wine or hard cider, and commensurate quantities of simple but delicious food. This can be cheap and a lot of fun.

Of course, you must choose the right place for your kind of people. Ornithologists, exhausted from a day's pursuit of the puffy pond piper (*Lacomelos inflatus* Ep.). may celebrate the discovery of two egg shells at the Rotten Rooster Inn. A group of hammer-happy paleontologists may relax at the Falling Rock Restaurant, after a smashing day in a quarry. In either case, the people will have dirty clothes and shoes, and probably a good time.

On the other hand, members of the Hollywood Association for Cheek and Chin Surgery (not to mention their Ladies' Auxiliary) may take offence when a lonesome butterfly drifts into the dining room. Put yourself in their shoes: in the afternoon, they have been to the dedication of a fully climatized center for nasal corrections, followed by visits to a winery, a private art collection and a monastery (with gift shops). Then, the insect reminds them of the unhygienic conditions so typical of Europe (didn't you know?); little wonder that the creature spoils the crowning end of a remarkable day, their eight-course black-tie dinner in a medieval French chateau. Different folks, different jokes! Just make sure that you entertain your people in the right style.

In places with a long history, there are rarely problems in designing excursions. The same goes for regions with beautiful scenery, especially if a boat trip, or a ride to a mountain top can be included. If a conference takes place in a city without historical sites, perhaps gardens, parks, a good zoo or aquarium, interesting modern buildings or monuments, museums, local craftshops (with demonstrations), technology exhibits, markets, breweries, wineries, or harbour roundtrips will all add up to an interesting trip. There are also options you had better avoid. They range from a trip on a polluted river with rotting warehouses and reeking refineries to a stop in a dangerous slum. Also it is utterly tasteless when people are bussed to reminders of war atrocities or concentration camps without being informed beforehand.

There are more potential problems with excursions. Often, poorly trained tour guides parrot a melange of nonsense and stupid jokes that the average tourist likes to hear. This may not go over well with scientists. Thus, you should ask a local expert to come along. When you negotiate with a tour operator, don't hesitate to bring up this topic.

Another, even more annoying, problem comes with the kickback system that pervades the tourist industry. Perhaps, you have wondered why your bus driver got very special treatment at a restaurant, or why he insisted that you must see this magnificent gift shop. Well, now you know. You may subscribe to the philosophy that there is nothing wrong when the oxen nibble while they are thrashing. Agreed. However, in some countries, the impudence of bus drivers and tour guides often exceeds tolerable limits. This can be very frustrating when, contrary to the agreement, bus after bus dumps your participants in front of a department store. While your people wander for 45 minutes through aisles loaded with junk, you realize that the buses will now get stuck in the rush hour traffic, which means that the evening program will have a poor turnout.

After this experience, I thought that I had seen it all. I was wrong. For the last excursion of the accompanists of a meeting, I had designed a special trip, including a lake with thousands of flamingos; prehistoric dolmens (structures built from gigantic stone slabs), probably contemporary with Stonehenge in England; a medieval town with remarkable churches, manors and a local museum; a magnificent mountain landscape with millions of wildflowers. This time, my bad luck started with the flamingos who did not come to their traditional breeding place. When I announced this, nobody dropped out from the excursion. Then came the day which will stick in my memory though I did not go along on the trip. First, the driver dropped the people outside the historic town and gave them just enough time to walk there and back to the bus. Next, he explained that the dolmens were nothing worth seeing and that he would not stop there. Objections by some people led nowhere. Then, he drove back over mountain roads that scared everybody into submission; however, the views were unforgettable for those who dared to look out. Only days later, did I learn the details of the affair. I am still grateful to the participants that they did not complain to me.

After this debacle, my recommendations are as follows: (1) Never sign a contract with a tour operator that does not spell out in detail: (*a*) what has to be done, and (*b*) what not. (2) Make sure the contract mentions that the tour operator will inform the bus driver(s) of all details of the contract, including tips. (3) For all-day excursions, stipulate that the tour operator will provide small maps that show the route and the stops. (4) Keep the initial downpayment so low that the agency must anticipate a severe financial loss if the contract is broken.

5

The program: how to accommodate
pigs in a poke

Logic suggests that in the daily program demanding presentations (especially major
lectures) should precede entertaining and interactive events. This translates to the
following order: Plenary Lectures, State-of-the-Art Lectures, Short Communications, Colloquia and Workshops, Poster Sessions, Forums, Business Meetings.
However, in practice modifications are usually necessary. During the first day of
the meeting, a welcome speech or Welcome Ceremony will precede the daily
program. On other days, special interest breakfasts may precede the sessions. Special
interest luncheons may be held during one or more noon breaks. In the afternoons,
Colloquia may be scheduled so that data shown in preceding Poster Sessions can
be discussed. A Closing Lecture will obviously be the last scientific event of the
meeting; however, it is debatable if it should be given before or after the Business
Session, if the latter is held during the last day of the meeting (see Section 3.2).

There are some frequently ignored rules:

The first one states: 'The more time for interactive events, the better.' Even at the
smallest conference, there should be a time slot for informal discussions
between participants. An afternoon consisting of four one-hour lectures followed by the immediate departure of the speakers is very unsatisfactory.

The second rule is equally important: 'Don't overload your program.' Unless local
(geographic, climatic or transportation) conditions make a different schedule
advisable, do not start the sessions before 9 am. At meetings of longer duration,
have lunch and evening breaks lasting at least 90 minutes. This is particularly
important at international meetings with participants from different time zones.

A third rule also expresses common sense: 'During parallel sessions, avoid overlap
of topics as much as possible.'

For meetings of several days' duration, there is a further rule: 'Adhere to a given
time frame as much as possible' (see the 'Standard Frame' given in Section
5.3 and Appendix B). When sessions are scheduled haphazardly, participants
may become confused and miss events.

5.1 Regional meetings

General regional meetings are usually conferences with fewer than 200 participants. If these have rather differing research interests, the organizer is challenged to make attendance worthwhile for as many people as possible. Obviously, interactive events are important, and the formal presentations must be carefully selected. The worst scenario is a regional meeting whose lectures are of interest only to a small minority of those present. This happens when the organizer selects speakers rather than topics.

At a regional meeting, the welcome/introductory remarks are usually followed by one or more Main Lectures of general interest. However, keep in mind that shorter lectures allow scheduling of more speakers. The afternoons are best used for Poster Sessions, and perhaps colloquia or workshops. If lectures are included in the afternoon program, the preferable time slots are the first ones following the lunch break. Forums or Business Sessions should be the last event of the afternoon. If a Forum *plus* a Business Session must be scheduled and two full afternoons are available, it may be preferable to hold the Forum on the first, and the Business Session on the second afternoon. However, try to keep the evening(s) free for informal interactive events.

During a two-day meeting, a social event in the evening of the first day is almost a must. The ideal setting is a buffet-style dinner or barbeque with drinks that allows participation of accompanists as well as informal discussions.

Overall, regional meetings vary greatly in style and length. Depending on local circumstances and habits, people may save money on accommodation by driving, perhaps for several hours, to the meeting. Keep this in mind and do not start too early in the morning. Consider in the design of the afternoon program that the same people may also leave early.

5.2 Small research conferences

It is easier to organize a small conference on a specific topic than a general regional meeting, provided a critical minimum of participants is available. The organizer does not have to worry about diverging interests and can custom-tailor the program according to the focus. However, here also, it is imperative to arrange a time for informal discussions, regardless of the length of the meeting. Though it is difficult to come up with a specific format for research conferences, the preceding section on regional meetings may give some ideas of how to arrange the events. Research conferences in a warm climate sometimes hold only morning and evening sessions, which allows the participants to spend the afternoons at a swimming pool or beach. This schedule will work only if enough people have a strong interest in the topics of the evening program, or no restaurants to which to go.

5.3 Major meetings

For practical purposes, one can divide large meetings into 'major meetings' with about 200–1000 participants, and 'mega meetings' whose attendance may reach 15 000–20 000. Mega meetings require a large, professional staff, and more time and effort than a full-time academician or researcher can afford. Their organization is not considered here.

Major meetings of longer duration usually accommodate a variety of events. For a meeting lasting four days or longer, these may be grouped and scheduled (using the international 24-hour clock) as in the following examples:

1. Early registration:
   Afternoon/evening preceding the meeting (16 : 00–21 : 00).

2. Special early events:
   (a) 8 : 00–12 : 00 Registration continued (on Day 1)
   (b) 8 : 00– 9 : 00 Registration continued (on Days 2–4)
   (c) 7 : 30– 9 : 00 Breakfasts for special groups (on Days 2–4)
   (d) 9 : 00– 9 : 25 Opening, welcome and announcements on Day 1.

3. 'Standard Frame':
     9 : 00– 9 : 45 Plenary Lecture
     9 : 45– 9 : 50 Break
     9 : 50–10 : 20 State-of-the-Art Lecture I
   10 : 20–10 : 25 Break
   10 : 25–10 : 55 State-of-the-Art Lecture II
   10 : 55–11 : 10 Coffee break
   11 : 10–11 : 40 State-of-the-Art Lecture III
   11 : 40–11 : 45 Break
   11 : 45–12 : 15 State-of-the-Art Lecture IV
   12 : 15–14 : 00 Lunch break (can be used for catered committee meetings)
   14 : 00–14 : 45 Plenary Lecture
   14 : 45–15 : 30 Poster preview (without presenters)
   15 : 30–17 : 00 Poster discussion (with presenters)
   17 : 00–18 : 40 Colloquia and/or Workshops (one or more in parallel)
   18 : 40–20 : 30 Evening break (participants dine at their own expense, unless Socratic Workshops are scheduled for the evening)

4. Evening events (20 : 30–23 : 30):
   (a) Informal get-together on the evening preceding the meeting
   (b) Welcome party on Day 1 (entertainment, beverages and snacks provided by organizer)

(*c*) Evening Workshops (Days 2 and 3)

(*d*) Fare-well party or banquet (Day 4).

5. Special events:

(*a*) Business or General Session

(*b*) Forum

(*c*) Closing Lecture

(*d*) Closing Ceremony, concluding remarks and/or thanks to the organizers

(*e*) Excursions.

Appendix B gives an example of how these events can be incorporated into a general schedule. Let's take a closer look at some of these events.

Registration Ideally, you would like to complete the registration before the informal get-together on the evening preceding the meeting. This way, you could participate in the party and arrange – hopefully minor – trouble shooting. Also, your staff would probably appreciate being invited to the party.

Most likely, this will not work out. During the first hours of registration, the organizer (or an experienced substitute) should be near the registration desk to handle critical situations, from the refusal of unacceptable checks to disputes with the hotel about fouled-up room reservations, and more unpleasant surprises, and most importantly, to collect the manuscripts for the proceedings (for more on manuscript collection, see Section 8.3.1). At major meetings, registration may go on well beyond the official time when lines form because many participants want to register on the way to the party, or a large group from a foreign country arrives on a late flight. Depending on the size of your meeting, it may therefore be advisable to have at least one *competent* person at the registration desk before and after the scheduled times on the day before the meeting. If you fear an onslaught of late arrivals during the first morning, extend the official registration also to include the lunch break.

Speaking of *competent* staff, at international meetings, fill critical positions only with people who are fluent in English!

Special breakfasts Some people like to use the early morning for committee meetings. Warn the chairperson that he may have a poor or belated turnout since, at conventions, many people stay up late. If he doesn't agree to switch his meeting to lunch time, just make a reservation as he insists. Next time, he will know better.

A senescent society may have a traditional breakfast for former presidents or other ex-officials. So, don't forget the toothless tigers. Just schedule their breakfast before the other events of the second or a later morning.

Opening exercises Veterans of scientific meetings may sigh when this or a similar term appears in the program. In order not to bore your participants with unnecessary formalities, keep the 'welcome,' including announcements, as short as possible; and schedule it so that it is kept within the time of the 'Standard Frame' (see next section). Prevent unnecessary speeches by government officials – especially if they did not give financial support for your meeting. I have suffered such meaningless sermons that filled the total time slot for the first Plenary Lecture. On the other hand, a Welcome Ceremony with some local flare (e.g., surrender of the keys to the town; native dances) can be entertaining for both participants and accompanists, if reasonably genuine. However, don't waste time for reasons of self-aggrandizement.

If possible, schedule the ceremony so that your scientific program can start with the first Plenary Lecture at 9 : 35. If this lecture is of the usual length (45 minutes), it will terminate at 10 : 20, which leaves you 5 minutes for a break before you continue with the first State-of-the-Art Lecture at 10 : 25. This brings your schedule into the 'Standard Frame' with the loss of one time slot for State-of-the-Art Lectures.

The 'Standard Frame'   When sessions begin at 9 : 00, the regular participants of your meeting can enjoy a leisurely breakfast after some early activities, e.g., jogging. Furthermore, if accompanists are present, they can be sent off so that participants do not miss part of the first lecture (see also Chapter 11).

With respect to morning sessions, consider the following. During three hours of lecture time, including short intermissions (5 minutes each) and a coffee break (15 minutes), you can schedule three Plenary Lectures (50 minutes each), five State-of-the-Art Lectures (30 minutes each), or seven Short Communications (20 minutes each). Alternatively, you can schedule four Plenary Lectures (45 minutes each), which would amount to a total time of 3 hours and 25 minutes; or you can combine two or more types of lectures. For example, one Plenary Lecture (45 minutes) and four State-of-the-Art Lectures, which would amount to a total of 3 hours and 15 minutes. This may be the optimal use of morning time (for a detailed breakdown, see Appendix B and Section 17.3).

Since it is customary not to have two Plenary Lectures simultaneously, you will perhaps decide to have one Plenary Lecture, followed by State-of-the-Art Lectures or Short Communications *in parallel*. This may be necessary at larger meetings, and you may need the advice of specialists. Wrong grouping of presentations is likely when organizers judge by the titles and do not read the abstracts. Often, parallel sessions are difficult to schedule, and it may be best to reduce their number as far as possible. As pointed out in Section 3.1.2, Poster Sessions are more fruitful than sessions with Short Communications. Just keep in mind that you can schedule 100 posters for the time slots required for seven Short Communications. Even if you schedule five Short Communications in parallel (total = 35), you waste the time for 65 additional presentations that could be given in a Poster Session of equal length.

Overall, the mornings are best used for lectures only, longer ones preceding shorter ones, and those with films scheduled last (see Section 3.1.5).

Afternoon sessions may begin with lectures, but should be used mainly for Poster Sessions and interactive events. My suggestion is to schedule a single Plenary Lecture immediately after the lunch break, i.e., usually at 14 : 00, and follow with a Poster Session lasting 2 hours 15 minutes (preview and discussion time combined). This will allow you to close the afternoon sessions with Colloquia or Workshops ending at about 18 : 40 (see Section 17.3).

Alternatively, you may schedule a Poster Session (3 hours), beginning at 14 : 00; and subsequently a Colloquium, Workshop or movies (1 hour 40 minutes). This way, you have more interactive events, and you will be finished at about 18 : 40. This schedule may be advisable when posters cover only a few areas of interest (see Section 3.1.2).

Evening events   Depending on the length of your meeting, you may have one or more evenings without social events. For these evenings, I recommend the Socratic Workshops as outlined in Section 3.1.4.3.

Business Sessions (or 'General Sessions' with similar agenda) are preferably held late in the afternoon; or, if time does not permit this, immediately after the evening break. Unless the issues raised require subsequent discussion among the participants before action can be taken (e.g., a ballot is cast), these events are best scheduled towards the end of the meeting when people feel saturated with science. A good time slot for Business Sessions is towards the end of a meeting in the late afternoon (17 : 00–18 : 40 time slot of the Standard Frame). A Business Session during the morning, when the mind is fresh, would be a mistake. If you wish to have a good turnout, schedule the session the day before the last day of the meeting, so that people who are to leave early can attend.

Forums   As in the case of business sessions, the best time for a forum is the late afternoon (17 : 00–18 : 40 time slot of the Standard Frame). Many Forums last 90 minutes to two hours; however, it is advisable to schedule them so that they can go overtime.

Closing Lecture   If a Closing Lecture cannot be avoided (see Section 3.1.1.6), it obviously has to be the last scientific presentation. Within the Standard Frame, the time slot would be the late afternoon of the last day, following the Poster Session. Otherwise, the last afternoon could be used for the Business Meeting, Closing Lecture and Closing Ceremony.

Closing Ceremony   It is a matter of taste how much time one wishes to spend on closing formalities. However, it must be kept in mind that, after several days of meeting, many people may not be in the mood to sit through a lengthy exchange of nice nothings. It appears preferable to keep the event short (no more than 15 minutes), or incorporate it as some toasts, etc., in the banquet, if the latter is held on the last evening of the meeting.

Excursions If a meeting lasts three days or less, no excursions may be indicated, unless there is general interest in visiting specific places. However, in biological, geological and related meetings, postmeeting excursions in the afternoon of the last meeting day, or on the following day(s) are usually the best option.

If no postmeeting excursions are planned and a meeting lasts four days or more, an excursion between the days of sessions may be indicated. During a four-day meeting, the afternoon of the third day is a good time. If a meeting has five full days of sessions, an all-day excursion (after the third day of sessions) may be called for. Since this will add another day to the meeting, it may be appropriate only if the majority of your participants is interested.

5.4 Summary and recommendations

From the preceding sections, it should be obvious that scheduling events for major scientific meetings is a true challenge. The main difficulty comes with the balance between lectures and interactive events. If the organizer subscribes to the philosophy that the primary purpose of his meeting is the exchange of ideas, he will try to reduce the number of formal lectures; especially Plenary Lectures of limited interest. In some societies, this may be a Gargantuan task (see also Section 3.1.1.1). My suggestions for major meetings are: (1) abandon Short Communications altogether; (2) keep the number of Plenary Lectures to a minimum; (3) carefully select State-of-the-Art Lectures; and (4) assign ample time to Poster Sessions, Colloquia and Workshops. On the other hand, at regional meetings with a more relaxed climate, provide an opportunity for younger participants to give brief talks, time permitting.

6

Selection of the meeting site: a touch of Russian roulette

6.1 General considerations

In the selection of a meeting place, the odds for a mishap are probably greater than in Russian roulette, and they grow with the naiveté and/or laziness of the organizer(s). To avoid major mistakes, it pays to make thorough inquiries about potential meeting sites. The more information you can get, the better. Ask both organizers and participants of recent meetings. Why also ask participants? Because organizers often remain unaware of serious flaws; and, on the other hand, they may be reluctant to admit major mistakes. Of course, the best recommendation for a meeting place is when it is used year after year by the same scientific societies.

Never trust a hotel or meeting facility without a written contract. If they refuse to sign one that is to your satisfaction, thank them for the warning and go elsewhere. Scientists typically totally underestimate the tricks of the convention trade. The example of a contract in Appendix I gives you some idea of what a skillful negotiator can obtain for a major meeting.

Whenever possible, prepare a list of questions and contact by phone the organizer of a previous conference at your envisioned meeting site. Perhaps, you can persuade him to send you copies of his contract(s). If his meeting included exhibitions, also ask about contracts with the decorator and exhibition service (see Section 3.4). Ask the managements of the hotel and meeting site for copies of contracts with previous organizations. Their reactions may be revealing.

Why is a written, legally binding contract so important? The following experience will answer that question.

When I organized a major international meeting, I made oral arrangements with the manager of the congress hotel. I had known him for years as a trustworthy gentleman, and thus was not worried about a contract. Unfortunately, he resigned from his job several weeks before our conference. The new management seemed very professional and cooperative. But then, at the meeting, we paid for my naiveté. To keep track of the number of participants and payments, my office had collected the accommodation forms and forwarded them to the hotel; thus we knew exactly which parties had booked for which of the two official hotels. Booked they had; however, when our participants arrived at the official hotels, many learned that they had been assigned to nearby hotels of supposedly equal standard. They were as surprised as I was; luckily for me, apparently nobody objected to the arrangement.

But soon came problem number one: how to get to the new hotels which were too close for taxicabs to bother, and too far for a walk with heavy luggage. I don't remember how we resolved the mess because, at this point, I was inundated with too many other worries.

Then came problem number two: people who had agreed to contact each other after their arrival at the congress hotel were unable to do so; the desk clerks, despite their computers, had no information on the fate of those who had been bumped from their reservations.

It took another day for problem number three to surface: people who had made room reservations starting in the late evening of the second day arrived, as scheduled. The desk clerk claimed to have no information on them, nor rooms available. I happened to witness the scene and resolved the situation. However, no good deed goes unpunished: I had a complimentary room at the hotel which I did not need. I gave it first to one bumpee, and then on the following day to another one. That was enough for an illustrious colleague to start the rumor that I was financially involved in the hotel.

Problem number four arrived as another surprise: breakfast at one hotel started 30 minutes later than in the other hotels; no big deal, except for the morning of an all-day excursion. The buses had been scheduled to pick up several hundred people from different places rather early, and a last-minute change of the departure times would have created monumental confusion. Aware of our predicament, the new hotel insisted that they could not provide an earlier breakfast unless – you guessed it – we would pay.

Why, you may ask, did the congress hotel bump our participants? Apparently, because they had contracts with travel agencies who would fill their rooms on a regular basis. Our meeting was just a one-time event organized by a novice.

6.1.1 Proximity of meeting site and accommodation

Ideally, one would like to have meeting rooms, overnight accommodation, restaurants, bar, parking and fitness facilities (including swimming pool and jogging trails) all in one safe, clean and reasonably priced hotel, with easy access to a railway station and/or airport. One may be willing to compromise and forget about easy access to railway and airport, provided the perfect meeting place is located in an attractive setting, such as a beautiful island or ancient town. Or, for financial reasons, one may consider a college campus during vacation time, provided it has a reasonable number of the above features.

There is, though, a frequent problem: meeting facilities located away from the overnight accommodation and/or restaurants. This may cause health-related and financial problems. Consider the following: foreign participants may arrive with jetlag, after many hours in airplane seats that would be considered too small for dogs. For them, and older travellers in general, it may be important to lie down for a little while during the day, especially after lunch. If they must take a taxi or shuttle bus to their hotel, they may not return to the afternoon sessions.

Long-distance travellers of all ages frequently experience gastro-intestinal problems, no matter how clean the food and water of the venue are. These people, especially when suffering from diarrhea, like to be close to their quarters.

Accommodation distant from the conference site may become much more expensive than anticipated. At a meeting in a major European city, the organizers did not check the recommended accommodation and left the hotel assignments to the local office of tourism – a cardinal mistake you should never make. The result was that they put some of us in motels about one hour from the venue by public transport. Of course, most participants took taxis, especially in the mornings; alas, the taxi fares were such that we would have saved money by renting rooms in three-star hotels close to the conference site.

The situation may be even more unpleasant when the meeting is held in an isolated hotel with lousy service and/or unreasonable prices. In this case, people will run off to other places, if possible (especially for dinner). If a reasonably-priced restaurant is too far away, they may buy provisions from the closest grocery shop or supermarket and eat them in their rooms. Of course, the experience will leave an indelible mark in their memory.

6.1.2 Location and accessibility

There is a simple rule: the smaller the meeting, the larger the choice of meeting sites. Mega-meetings with thousands of participants can be held only in a limited number of cities, and reservations may have to be made years in advance. Smaller conferences can convene in many places, which allows the organizer much flexibility. If your participants (and accompanists) enjoy a beautiful landscape, select a nice beach, mountain or castle; if they are mostly interested in cultural events and/ or shopping, select a large but safe city. However, avoid high mountains; even those without obvious heart conditions may suffer (especially during the first night) when they are lodged at places at more than 2000 meters above sea level. If your participants anticipate swimming outdoors, make sure that the pool will be open (heated, if necessary) during the off-season; also, remember that on some ocean beaches high waves, poisonous jellyfish or unfriendly sharks do not permit swimming during certain times of the year. Last but not least, make sure that your location can be reached, at any time, in case of a medical emergency. This excludes quite a few mountain resorts during the winter season.

6.1.3 Safety

There are many places that would be ideal for conventions if they were safe. For example, one ancient capital is disqualified because, during rush hours, there are too many hands in the pockets of foreign males, and all over the body of foreign females. Another ancient city cannot cope with its robbers on motorcycles. These fellows rip pocketbooks from female tourists, or smash the backwindows of stopped cars and grab within seconds everything they can (which often includes money and passports).

In one beautiful tropical capital with excellent meeting facilities, street crime was so rampant that foreigners were advised not to leave their hotels. Here, in front of my hotel, a taxi driver grabbed a bank note from my hand and dashed off. Minutes later, at the same place, several female tourists were beaten and robbed. I wonder what happened when an international feminists' convention was held there a week later.

In some places, crime peaks during the warm season, and again before Christmas. Thus, the delegates of a convention held in January may have very different experiences from those of delegates at a convention in July. At any rate, it is not pleasant to stay in a hotel that is hermetically sealed, by armed guards, from the local population, be this in an industrial city or on an exotic beach.

There is no need to extend this list. The sad fact is that crime has become a factor in the planning of conferences. If your meeting site is not safe, be honest with your participants and warn them in no uncertain terms of the local problems. At one recent meeting, the organizer included a map of the city, with dangerous zones clearly marked, in the program. This openness was certainly appreciated.

6.1.4 Medical care

During a major meeting, around-the-clock medical care must be available. When you discuss this issue with the hotel management, don't let them fool you: get in writing that a physician will be on the premises or nearby, not just from noon to 5 pm, but all the time; and make it clear that this information will be printed in the program. Apart from routine gut problems, many serious situations can arise. Examples I have had to deal with include: malaria-like fever of uncertain nature, lasting for days; uncontrollable asthma attacks, lasting for hours; dangerous inflammation of a wound caused by a sea urchin; food poisoning; heart attack late in the evening; disappearance of a special foreign insulin before the critical morning injection.

6.1.5 Baby sitting service

If you anticipate a large number of parents with children at your conference, try to arrange baby sitting services. Any major congress hotel should have pertinent information, including (a) the ages accepted, (b) times for which the service is available, and (c) daily or hourly rates. Provide this information in the mailings that go out with the critical preliminary program (see Section 17.2).

6.1.6 Advance payments

Sometimes, meeting facilities demand a deposit years in advance. Forget these, even if you have funds available. You cannot predict what will happen to your conference. Especially with international meetings, an economic, financial or political disaster may force you to curtail or cancel the event.

6.2 Selection of hotels and meeting facilities

6.2.1 Obvious criteria

If you can't get reliable information on an envisioned meeting place, good sleuth work is indicated. Don't trust the number of stars: a four-star hotel in one country may rate as a two-star hotel in another one. And don't let looks deceive you. I recently stayed in a beach hotel with a magnificent garden and swimming pool that is highly rated by two leading travel guides. Apparently, their investigators never slept in the place; otherwise, they would have discovered that, on top of other deficiencies, the toilets inevitably clog because of a flaw in the design.

6.2.1.1 Accommodation
For hotels and motels, I suggest the following checklist:

Hallways Are the carpets and walls clean, the floors and walls without cracks? Are all ceiling panels in place? Is the furniture in good shape? Note: spider's webs and missing light bulbs are bad signs.

Elevators Are they clean, with an emergency phone or similar device? Are there sufficient elevators? In one major congress hotel, we had to wait up to ten minutes during the rush hour!

Guest rooms Are the answers to the following questions satisfactory?

(1) Are all bedrooms of a given price class comparable? In a small hotel ('pension') in a central European town, I paid more for my room than I would have in a large hotel in New York City. To add insult to injury, my room

was half the size of other ones for which the same exorbitant rates had been approved by the local authorities. Who made me aware of the situation? The maid who claimed the owner even stole her tips!

(2)   Are the bathrooms clean and functioning, with showers (with hot water)? Do you smell odors from the sewage system? How about dirt or grease on lamp shades, the tops of cabinets and the carpet under the bed? Particularly in warm climates, check for cockroaches (e.g., in dark corners of cabinets or drawers, and behind pictures). And of course, are the bedsheets okay? In a hotel for visitors of government agencies, I found human hair (not mine) between the sheets – you get the picture.

(3)   Do the door locks work? Once, I was trapped in my room in a dormitory long enough to miss an all-day excursion.

(4)   Do the electrical outlets and lamps work? Some men get upset when they can't use their electric shaver in front of a mirror; and there are people who need to read in bed so they can fall asleep.

(5)   If present, do the telephone and TV work?

(6)   Does air conditioning and/or heating work? Also, ask if they can be used if unseasonal weather strikes.

(7)   Are you satisfied with the following indicators of hotel quality:
Chair(s) and desk for writing; complimentary stationery; local information (usually a weekly or monthly publication); sufficient hangers for clothes; sufficient towels; complimentary things like soap and shampoo; contents of the room bar (if present)?

Special services   Are vending machines and ice makers present, and working? Is there a parlor or a machine for shoe shine? Do they have a fast laundry service? Can faxes be received? Do they have Internet connections?

6.2.1.2 Meeting rooms and equipment

When scientific events are held in parallel, make sure that the distance between meeting rooms is short. Do not hold parallel sessions (*a*) in two different hotels, and (*b*) on a campus where auditoriums in different buildings must be used. I remember a meeting where an icy street prevented participants from attending sessions in a nearby hotel. At another meeting, the distance between campus buildings induced us to make a shorter trip, namely to a nearby pub.

The following questions summarize important initial criteria. When you discuss them with the management of prospective meeting facilities, don't let them dupe you with wishy-washy answers.

(1)   Are there enough rooms of the right size(s), and will they be available at the time of your meeting? Be suspicious when a local club has been using one

of the envisioned rooms routinely during a time when you will need it for
sessions.

(2)   What is the lawful maximum number of occupants, and are there enough
      seats in the meeting rooms? There is an old trick: they will show you one
      room set up with chairs, but not tell you that they don't have enough seats
      for the other rooms. Also, poorly designed seats can be worse than no seats
      at all.

(3)   Are the seating arrangements satisfactory? Do you need these to be 'class
      room' style (seats and tables), or 'theater style' (chairs only)?

(4)   Are the rooms acoustically well isolated from each other? Is there interference
      from the outside; e.g., from traffic? If moveable partitions are used to separate
      rooms, check carefully if they really work. In my personal experience, they
      never do! It can be very unpleasant when people use hammers to put up
      posters while in the neighboring room speaker and discussants try to outtalk
      the noise.

(5)   Are the rooms well ventilated? Often, smaller rooms are so poorly ventilated
      that they need air conditioning even on cool days.

(6)   Are the rooms supplied with heating and/or air conditioning? Even if so, ask
      the management to show you that the noise of these systems is tolerable for
      people in the backrows.

(7)   Are there projection screens and, if needed, blackboards? Are both large
      enough? Is all projection equipment (including *wireless* microphones and
      pointers) present, or locally available? If it has to be rented, check out the
      place and its prices. Avoid renting laser pointers; in the hand of nervous
      persons, they can become a torture for the audience. If needed, do the
      monitor(s) and audiovisual system work?

(8)   Can the rooms be darkened sufficiently for slide projections and films? If
      not, you probably can't use them for scientific presentations.

(9)   Can slides without a glass cover be projected, or is the room so large that
      powerful, heat-generating projectors must be used? In the latter case, even a
      warning in the preliminary program is unlikely to prevent some participants
      from arriving with unprotected slides that may melt during projection.

(10) Is the room for poster sessions of sufficient size, well illuminated, and do
      they have enough and adequate poster boards? For details, see Section 3.1.2.

(11) Are there clean toilets, separate for each gender, near the meeting rooms?

6.2.1.3 Gastronomy

Restaurants are critical for the success of a meeting. If the food is bad, the prices
outrageous or the service unfriendly, it will strongly interfere with the happiness
of the participants. As pointed out above (Section 6.1.1), one of the worst scenarios
is unsatisfactory food service at a meeting place from which participants can't
escape to better restaurants.

The best way to serve breakfast and lunch is as buffets where people can make their own selections. A meeting hotel should be able to offer you bargains for your participants, in which case advance sale of coupons may be the most efficient arrangement. Another important factor is the availability of alcoholic beverages. In the United Kingdom and some of its former colonies, restaurants need a liquor license. Make sure there are no problems with this; participants of your meeting may not be in the mood to run to a 'bottle shop,' especially before meals. Last but not least, don't forget to check the toilets. Are they separate for each sex? Are they clean? How about towels or warm-air driers? Are the wash basins inside the toilets?

6.2.2 Subtle criteria

Assuming you found the accommodation, meeting facilities and restaurant satisfactory, continue your inspection using subtle criteria. Here are some suggestions.

(1) Unless safe deposit boxes are built-in in all rooms, ask at the reception desk whether they are available elsewhere. If they are not, chances for thefts grow with the size of your meeting. If they have special safe deposit boxes, ask how much they charge for one. An unreasonable price may tell you all you want to know about the management.

(2) Ask them if they can make a few photocopies for you. If they can't, or charge you too much – that's a red flag.

(3) If you are running an international meeting, find out their currency exchange rates. If it is worse than those of other hotels, beware! If they are more expensive than the local banks or other places of exchange, ask them, like a naive tourist, about the exchange rate outside the hotel. If they claim their rate is better, they are both greedy and stupid. After all, you would probably discover the truth on your first stroll down the street.

(4) Check the attitude of the employees at the reception desk. Ask them for help with local arrangements: a ticket purchase; finding a special restaurant and making reservations there. If taxis are not routinely available, ask them to call one. The reaction to your requests will tell you if they are well trained, which in turn tells you how the place is managed.

(5) Have someone come to the reception desk late in the evening and ask for you. First, with a misspelling of your name; and if that doesn't work, with the correct spelling. If the employees can't find you with a little misspelling, it's a bad sign since names are often misspelled. If they can't find you with the correct spelling, definitely choose another place for your meeting (see also Section 18.1). Like other tricks, I learned this one the hard way: at an international meeting, people arriving late at night were denied their reserved beds because the receptionists in four (!) hotels were unable to use their computers. Don't let the smart uniform of desk clerks distract you from their lack of training.

(6) Make a long-distance call, or even better, an overseas call from your room. Some hotels make obscene charges, a multiple of what you would pay at a public phone nearby. In one congress hotel, the cashier warned us not to make overseas calls because he was sick of fighting with people who almost dropped dead when he presented their telephone bills.

(7) Have someone send you a fax, to be picked up at the front desk. See how they handle the matter.

(8) See if they charge for chairs in the swimming pool area. If they do, they may come up with more surprises. This is not the kind of outfit you want to deal with.

(9) Find out: (a) what the parking fees for your participants would be, and (b) whether the parking facilities are secure. Unreasonable parking fees and/or unguarded parking lots invite trouble, from complaints about prices to break-ins and stolen cars.

6.3 Special considerations at international meetings

6.3.1 The political situation

If a country is politically unstable, don't even think about holding a meeting there. The prospect of personal danger may keep many people away, no matter how attractive your program is. Furthermore, bad news could cause last-minute cancellations that would make the program a shambles, and your budget a disaster.

6.3.2 Immigration policies

Scientific meetings should never be held in countries that discriminate against persons because of their national origin, race or religion. It is also inadvisable to have a conference in a country with a capricious visa policy. I remember a research conference at which several foreigners arrived days late because of visa shenanigans, while others never showed up at all. If you plan to host an international meeting, carefully check the latest immigration rules of your host country.

Recently, the organizers of a research conference were shocked when a scheduled speaker, after thousands of miles of air travel, was turned back at the airport of their capital. How could this happen? She came from a nation that had become a fashionable target of boycott, in which the host country of the meeting had just joined.

6.3.3 Attitude towards foreigners

It is bad when travelers are routinely humiliated at the border. It is even worse when religious dogma, in the form of dress codes or restrictions on food and drinks, is forced upon visiting foreigners. Often, uneducated persons consider people unfamiliar with their language inferior, and treat them accordingly. In one country, this causes almost regular cycles in which years of rotten treatment of foreigners are followed by years of economic hardship because travel agencies don't dare to send tourists. In a case like this, don't take chances.

In the same country, we had an unforgettable experience with the shipment of programs and abstracts. A young woman, representing the firm who was to deliver them after their arrival by airplane, did not bother to inform her boss that a customs official was neither willing to check our shipment nor talk to me, a foreigner (even though I speak the language). The miserable bureaucrat was finally persuaded to release the material four days later – during the first morning of sessions, and after payment of an obscene customs fee. In this case, the ego trip of a little government employee had almost caused a disaster for a large international meeting.

6.3.4 Financial stability

You cannot come up with a reasonable budget projection when a country is notorious for unpredictable currency exchange rates. Even in countries with a more stable currency, there can be surprises when hotels charge payments in dollars rather than their own valuta. For example, when the participants of one meeting checked out, they were told by the cashier that they owed more money because their payments had been made several months in advance; i.e., when the exchange rate for the dollar was different. The furious foreigners argued that the hotel had been working with their money for quite a while, and that they had no funds left. They were kindly asked to use their credit cards, which they finally did in order to spare the gracious organizer of the meeting (who, we hope, never learned about the robbery).

6.3.5 Strikes

Some countries seem to be in a continuous state of variable paralysis due to 'industrial action.' Do not hold meetings in a country where air controllers, railway personnel, the postal services or hotel employees have a history of frequent strikes. Just imagine the situation if your participants do not arrive at all, or arrive days late, at the meeting. How do you handle refunds after spending much of the registration fees for services (e.g., secretaries, telephone, printing of program and abstracts, deposits for social events) when your calculations were based, say, on fees from 300 more participants? What do you do when your participants are left stranded in their hotels without funds after a meeting?

7

The dates of the meeting: you can't win

The best times for scientific meetings are probably the pre- and post-seasons. The advantages are obvious: reasonably good weather, no mass tourism, reduced room rates, and frequently lower airfares. In most of Europe, weather conditions make late spring and early autumn equally attractive; in the southeastern United States and the Caribbean region, on the other hand, the hurricane season (from about August to December) is a risk factor for larger meetings. Similar considerations apply to many places in southern and eastern Asia with seasonal typhoons. Also, it will not create fond memories when your participants grow mildew on their heads while waiting for the repair of a bridge during the monsoon. Of course, it also does not make sense to select locations where snow or ice could prevent participants from either arriving or leaving. Will you pay for their rooms when they are trapped for days in an expensive airport hotel? Furthermore, don't choose a time when many families traditionally get together, i.e., in particular between Christmas and New Year. Last but not least, remember that air fares may be extremely high at the weekend. This could be a deterrent for prospective participants when a meeting closes on a Friday or Saturday.

There is one more factor to consider: special local events. No matter what type of meeting you envision, make sure that your meeting does not clash with an event that causes local overcrowding of roads, parking lots, restaurants, hotels, etc. Typical examples would be major conventions or sports events.

When preparing for international meetings, religious practices must be considered. Even if a Muslim country is considered 'tolerant,' it would not be advisable to hold a conference during the Ramadan. There is always a chance that feasting and drinking infidels would incense believers. Similarly, devout Christians may refuse to attend meetings that take place during Easter or Pentecost holidays. On one occasion, we had to revise a meeting program when it turned out that some colleagues wished to observe Sabbath.

In the case of international meetings, differing teaching schedules are another problem. For instance, European universities (with a typical summer semester) may teach during June and July, the traditional vacation time of most North American universities. Summer vacations in the southern hemisphere usually coincide with the winter semester of universities in the northern hemisphere. Since there seems to be no time that suits everybody, the organizer of a major international meeting is likely to make some people unhappy. My advice would be to pick the best time for the majority of prospective participants and announce the meeting well in advance. At least that gives colleagues a chance to make special arrangements, if necessary.

Finally, why are so many meetings held at the wrong time? The answer is: money. Often, managers of hotels and meeting facilities go out of their way to lure conventions for the worst time of the off-season. Don't fall into that trap, and return their beautiful presents immediately. Your reputation is worth more than six bottles of champagne and free rooms for your family during the summer vacation.

8

Publications: cruel and unusual punishment

8.1 Weighing the pros and cons

One of the earliest decisions to be made concerns the publications of your meeting. An abstract volume is a must in most cases. Proceedings are a different question. Consult the *Index to Scientific & Technical Proceedings* (Institute for Scientific Information, Philadelphia, PA, USA) to confirm that your envisioned proceedings will not overlap with similar recent publications. Since there are enough unnecessary papers hampering scientific progress, don't have proceedings unless there is a good reason for them. Remember that one of the most fruitful series scientific meetings, the Gordon Conferences, do not result in any publications. If you are running a smaller research conference, it may not even pay to have abstracts.

Consider the following questions:

(1) Should there be any publications at all?
(2) If you decide to have abstracts:
   (*a*) Should they appear in a journal (thus becoming quotable publications), or should they be compiled in an abstract booklet just for perusal at the meeting?
   (*b*) Will they be available at the meeting, if published in a journal; or would you need an abstract booklet, anyway, since the journal issue could not appear before the meeting?
   (*c*) Which journal(s) should be considered and contacted immediately?
   (*d*) Should you simply make them available electronically on the World Wide Web?
(3) If you decide to have proceedings:
   (*a*) Which contributions should be published (all lectures, and the abstracts of posters; Plenary Lectures and abstracts of Short Communications; or just the Plenary Lectures)?
   (*b*) Will it be possible to enforce submission of the contributions by an early deadline? See Section 8.3.1 for more on this topic.
   (*c*) Which publisher(s) should be contacted?
   (*d*) Do you want or need coeditors?

Assuming you have decided tentatively to have both abstracts and proceedings, let's consider some important details.

8.2 Abstracts

Abstracts can be meaningful and worth quoting; perhaps more often, though, they should be forgotten as fast as possible. Apart from the content, the deadline for the submission affects the quality of abstracts. When submitted too early, abstracts contain vague and half-fermented information. If the deadline is late, some may not arrive in time for printing. In short, the deadline for abstracts creates one of those no-win situations for the organizer.

For a major meeting, the ideal schedule would be perhaps as follows: (*a*) deadline for submission: six weeks before the meeting; (*b*) compilation, page numbering, indexing and clerical work, including retyping of damaged or poorly prepared abstracts, immediately following the deadline; (*c*) delivery to the printer: five weeks before the meeting; (*d*) return from the printer within one week; (*e*) immediate mailing of the abstracts, together with the program, to the registered participants (by air mail for overseas participants).

Of course, this is a dream. On rare occasions, it may come true, provided: (*a*) you can rigorously stick to your deadlines; (*b*) you have excellent assistance; (*c*) the printshop is reliable; (*d*) the postal services are better than experience has led us to expect; (*e*) you can afford the postage.

The reality is usually different. For a major meeting, a reasonable deadline for submission of the abstracts may be four months before the conference (see also Section 16.4). If the abstracts are to appear as a journal issue, you may have to set the deadline earlier. Also, regardless of whether they will appear as a booklet or a journal issue, sufficient numbers must be available for late registrants at the time of the meeting. If the meeting site is distant from the printshop, this raises the question of whether they should be hand-carried or should you risk mailing them? Also remember that during vacation time, your participants may have already left on premeeting travels when your abstracts arrive only days or a few weeks before the meeting. For minor meetings, a deadline one month before the conference may suffice if you plan to hand out the abstracts at the registration desk. As with all deadlines, in case of doubt set an earlier date.

For reasons of accuracy and speed, publication of the abstracts in camera-ready form is preferable; suggestions for the preparation of abstracts and abstract forms are given in Appendix J. If the abstracts have to be reset for printing, a new source of potential errors is introduced. Furthermore, proof-reading will be necessary, which means more work. Hence, you might consider accepting electronic submission of abstracts.

When abstracts for different types of events are expected (e.g., for Main Lectures,

Short Communications and Posters), make sure that you have clearly marked, special forms for each type of abstract. If you send out only one form, with boxes for the different types of abstracts to be checked by the authors, people may inadvertently or purposely mark the wrong box. Why should someone purposely mark the wrong box? Because he wants to give a lecture, and not the poster for which he was invited. Unlikely? Not at all. Take it from an organizer who has had to fight it out with an 'imposter.'

For *abstracts of talks or Posters*, often the rules for abstracts of research papers apply (see, e.g., R. A. Day, *How to Write & Publish a Scientific Paper*, 4th edition. Oryx Press, and Cambridge University Press; 1994). However, up to three references (without titles, only the name of the first author spelled out) should be added. For clarity, it may be preferable to give the references after the text, but this is not an iron-clad rule. The references will be helpful for colleagues who are interested, but not familiar with the work. For a good example, see the abstracts in the *Journal of Physiology (London)*. *Prima facie*, an abstract with references has more credibility and should be considered a quotable publication. Abstracts without references, especially when ending with the notorious '. . . will be discussed,' do not necessarily instill confidence. Often, they result from a submission deadline that was too early (see Chapter 16), and/or from data the author hopes to have at the time of the meeting.

Abstracts of reviews must be custom-tailored to suit the topic. They should contain clearly stated conclusions and, whenever applicable to the topic, point out 'where to go from here.' Meaningful abstracts of reviews are difficult to write. They require a careful selection of highlights or representative examples. Sometimes, the phrase '. . . are discussed' may be unavoidable. If it appears twice, it should indicate a sloppy preparation and/or lack of knowledge.

The road to an abstract issue may give you a foretaste of what to expect with the proceedings. No matter how hard you try, some people will probably submit sloppy, unacceptable abstracts. Usually, the same people do this meeting after meeting; and often, they are also notorious manuscript delinquents (see Section 8.3.6). In some instances, you may simply not invite them; or you may forewarn them person-to-person that their abstracts will be refused if of unacceptable quality. Unfortunately, this is not always feasible, which means that their abstracts may have to be retyped by your staff. Perhaps, you may have to accept this as a fact of life. Be this as it may, if these people submit manuscripts for your proceedings, you are likely to have problems.

An abstract volume does not have to be expensive. A good printer can do a perfect job with a copy machine and an automated, or semi-automated, binding device. Some societies have superbly printed abstract booklets on expensive glossy paper. Usually, these booklets also contain advertisements which paid for most or all of the costs. However, someone has to solicit the advertisements. If your meeting is not of particular interest to commercial sponsors (for example, manufacturers or suppliers of laboratory instruments or chemicals; pharmaceutical firms; local restaurants or shopping centers), don't waste your time on a wild goose chase. Just make sure that your abstract volume is well designed and printed at a reasonable price.

Of course, most of these worries are taken care of when an *experienced* person (e.g., the executive officer of a society) routinely handles compilation, indexing and forwarding of the abstracts to a journal; and then makes sure that they are mailed to the participants. This is the job for unsung heroes whose work is rarely appreciated.

8.3 Proceedings

8.3.1 Once more: should you really go ahead?

Before deciding on the publication of proceedings, think of the young Napoleon who declared that he would never take chances if he could avoid it. Remember what happened when he later forgot his resolution. If you cherish your peace of mind, proceedings are not for you. Once the contract with a publisher has been signed, you are assured a lot of action, and for quite a while, you will not know if your road leads to Austerlitz or Waterloo.

Perhaps, the following will help you decide:

(1) Will the expected contributions be worth publication? In other words, would you buy the proceedings if you were not involved in the meeting? Or would you just get photocopies of a few papers because: (*a*) the quality of the contributions varies; or (*b*) the contents of the proceedings are heterogeneous and contain only a few papers of interest to you?

(2) Will you be able to avoid notorious manuscript delinquents when you select your speakers? An outstanding meeting without proceedings may be preferable to a conference whose lectures will be published two years later.

(3) In the case of a larger number of contributions, can you readily afford, time, money and otherwise, the endless reminders by phone, fax, Internet, e-mail or express mail? It is unlikely that you will weed out all manuscript delinquents in advance, even if you avoid the notorious ones. For more on this problem, see Section 8.3.2.

(4) If the proceedings are to be published by the camera-ready process, can you afford to have many, if not all, manuscripts retyped? Despite your carefully prepared instructions, the chances are you will receive only a few perfect manuscripts. Since returning the poor manuscripts to the authors would create another wave of problems, you will need a secretary who is (*a*) a good retypist, and (*b*) a good proof-reader; that means, someone of a kind that is

rare and usually expensive. And don't forget, someone – not a butterfingers – has to transfer the figures to the new manuscripts! This will be particularly difficult when both text and figures have been printed by a laser jet or similar device on the same page.

If the publisher requests submission of computer diskettes, pray that all of them can be used. It is very important to give your contributors specific technical and labeling instructions.

(5)   Should you publish the proceedings (a) as a book, (b) as a special issue of a journal, or (c) as groups of papers in one or more journals?

(6)   Will there be sufficient interest in the proceedings to make their publication financially viable? Try to come up with a realistic number of guaranteed sales, based on the assumptions that: (a) all senior authors will receive free copies of the proceedings; (b) the other participants of your meeting will buy them at 50% of the regular sales price; (c) you will be able to include the reduced price of the proceedings in the registration fees (see also Chapter 13). Very few books of this kind sell more than 500 copies.

(7)   With these figures on hand, can you find a publisher who is interested in your proceedings? If you are declined by several experienced editors, it may not be worth pursuing the matter further. For more on the selection of a publisher, see Section 8.3.5.

(8)   Will it be necessary or desirable to have the proceedings reviewed? For a detailed discussion, see below (Section 8.3.3).

If you decide to have proceedings, don't make the classic mistake of many organizers: delayed action. An early decision, followed by an agreement with a publisher, is essential. If you inform your speakers only a few weeks before the meeting that manuscripts have to be handed in at the registration desk, you are committing a cardinal blunder. Some people will never deliver a manuscript; others will give you some rehash of an earlier paper; and others still may give you something they couldn't publish elsewhere. Many manuscripts will be submitted after the meeting, some in bad shape.

Mailing early information is not enough. Be as precise as possible when explaining what you expect: scope (review or original data); number of pages (include spacing and margins) and words; details on the publishing process (camera-ready or regular printing); type of illustrations (photographs or line drawings only); precise format of references; and of course: the deadline. Ideally, these details should be conveyed to the contributors about one year before the manuscripts are due. Often, though, six months may be the best you can do.

8.3.2 Deadlines and dead lines

The attitude towards manuscript deadlines tells you a lot about your peers. More precisely, it is the excuses they come up with when they miss a deadline, or ask for an extension. Some are very imaginative, others just honest. Others still can't have changed their story since they missed an assignment in kindergarten. With certain personalities, appeals to decency (you paid the character all travel expenses in advance, plus a lavish allowance) or responsibility (he is holding up the publication of proceedings for which hundreds of people have paid months ago) are totally wasted. In such cases, very firm action is needed. See the slightly modified version of a letter (Appendix K) that worked reasonably well.

There seems to be no correlation with rank, race, religion, age or experience. Illustrious senior scientists and well-known editors may try your patience with the same excuses as graduate students. Not rarely, people who went to any length to be invited speakers turn out to be the worst manuscript delinquents. In my experience, women tend to be more reliable than men, but I hesitate to subject this notion to a statistical analysis!

Thus we come to the question: How can you enforce manuscript deadlines?

First of all, set an optimal date. Usually, this will be at the very moment when people pick up their portfolios at the registration desk. Have an assistant ask the authors for their manuscripts and keep a tally. This procedure will enable you to identify and contact manuscript delinquents early in the meeting. On the other hand, the worst deadline for manuscript delivery would be after a meeting, especially before the summer vacations. What will you do if manuscript delinquents cannot be contacted for many weeks?

Second, reinforce at every opportunity (e.g., mailing of information relating to your meeting) that the manuscript deadline is meant to be taken seriously.

Third, apply the strongest weapon of all: money. No travel support until an *acceptable* manuscript is in your hands! Spell out what is meant by 'acceptable': a complete manuscript with publishable figures (I have been stuck with drafts, papers without references and hand-drawn graphs!). Assuming that the manuscripts are due at the time of the meeting, inform the authors well in advance that the manuscripts will be read upon delivery, then you will decide on the financial support.

If an acceptable manuscript was handed in on time, and the author then implores you for an extension for a revised version, keep the manuscript you have and set a date: tell him that if the new deadline is passed, the manuscript you have will be published as is.

Even if you are unable to give travel support, you may yet have some financial clout: make sure that price of the proceedings is included in the registration fees for *all* participants. If the publisher provides free copies for contributors, announce that a refund will be given only to authors who submitted acceptable manuscripts (or proofs).

8.3.3 Reviews: are they worth the pain?

The editor of proceedings will always be under pressure. In dynamic disciplines, proceedings are often outdated within two years. Hence it may be preferable to sacrifice lengthy reviewing procedures in favor of fast publication. After all, the term 'reviewed publication' has an appealing ring for the naive rather than the realistic (for a recent discussion of the issue, see P. McCarthy in *The Scientist* **8** pp. 1 and 21; 1994). Let's face it: (1) in this day and age of specialization, you may not find competent reviewers for certain contributions; (2) older scientists may agree that over the past two decades, the relative decline in research funds has been accompanied by an increasing number of meaningless, often unfair reviews; (3) some people are so desperate to get published that they will comply with the demands of reviewers, no matter how asinine they are. If you invite good people, the chances are that their manuscripts can be accepted as submitted. On the other hand, papers from people who do not publish in good journals are unlikely to improve your proceedings.

When proceedings appear as an issue of a journal, reviews may be required. To expedite the reviewing process, a clear agreement with the editor of the journal – well in advance of your meeting – should cover the following: (1) format and scope of the submissions; (2) reviewing procedures, such as the required number of reviewers per paper; (3) deadlines.

When your proceedings are published as a book, you may not have to send out the manuscripts for reviews. However, you and your coeditors must check the papers for obvious shortcomings. In critical cases, you may yet consult a specialist. Whether this procedure justifies the term 'reviewed' is a matter for debate. For instance, if each manuscript is read by you and two coeditors, and the conference was in the field of specialization of all three of you, I would call it 'reviewed.'

To speed up external reviews, it is essential to organize a group of reviewers before the meeting. If these people attend the conference, give them copies of the manuscripts so they can start working right away. However, always keep the original, especially when it contains photographs.

It makes sense to ease the burden of the reviewers as much as possible – especially when you expect them to pay overseas airmail from their own funds! Here are some suggestions:

(1)   Design (or use) a form that has all the needed information for the reviewer.
(2)   Make sure that the form can be used easily on a computer. Neither secretaries nor scientists (who type themselves) like to waste time juggling with the format of a poorly designed form page.
(3)   Have the name(s) of the author(s), the title of the paper, and the name of the reviewer typed on the form before you send it out.
(4)   Enclose an envelope with your address, and if possible, with postage.

8.3.4 Selection of coeditors

Do you need coeditors? If you are experienced and the proceedings will be a smaller publication, certainly not. If you are inexperienced, it would be better to team up with someone with a proven record, even if you do not anticipate a major volume.

Coeditors can be helpful, provided: (*a*) they are reasonable and reliable; and (*b*) the individual chores are clearly understood, right from the beginning. Otherwise, you are asking for trouble. For example, the proceedings of one symposium appeared two years late because three fellows could not agree on who should be the 'senior' editor.

On the other hand, two or even three editors can speed up a major publication if, for example, they agree on the following division of labor: Number one negotiates with the publisher, designs the general layout of the proceedings, collects the manuscripts and goes through all related pains with the authors. Number two edits the manuscripts (which means careful reading and supervision of the retyping). Number three prepares the index and thus also double-checks the manuscripts.

In case of doubt, an experienced editor can handle all of the above by himself if he anticipates problems with potential coeditors. He will also try to avoid transoceanic coeditors since this creates delays and added expenses.

8.3.5 Selection of a publisher

It is not always easy to find a good and reliable publisher. Contrary to widespread belief, the size of a company is no guarantee for quality work and good marketing, or cooperation. Big publishing houses sometimes have bewildering infrastructures and a peculiar decision process.

Should you really worry how your publisher runs his shop? You bet! To minimize the time between submission of the proceedings and the date of publication, you need a publisher who has: (1) an experienced and understanding staff; and (2) sufficient capacity to squeeze in your – probably belated – manuscripts.

When a publisher insists on proof-reading by authors, be sure to check that the publisher has the ability to send out all the papers by express mail or courier service. The publisher should inform all authors in advance of the date on which proofs will be sent out. The letter of warning should state that unless corrections are received by fax, express mail or e-mail within five days then the publisher will assume the author has no corrections.

You also need a publisher whose staff can give authoritative answers, for example, on the format of the index, unusual references (e.g., how to abbreviate 'bibliographic tapeworms' with twelve authors, and how to quote confusing handbook citations), and potential problems with figures. When the camera-ready process or electronic files are used, there must be especially clear instructions: Which font(s) can be used? How are Latin species names to be typed? What has to be underlined? Are italics required? What kind of figures and tables can be submitted? If you are referred to different departments and/or different people for these questions, you may get different answers. This should be a warning signal.

In the early days of camera-ready printing, my coeditors and I had to deal with three different editors of a publishing house. As a result, we had to spend hundreds of dollars on retyping and the fine art of mortising (cutting holes into pages and replacing lines).

Books may be fountains of wisdom, and forever; but you can't expect this automatically of a publisher's staff. Here are some examples why from my personal experiences:

You have reached a verbal agreement with an editor. Suddenly, he/she leaves and the successor has different ideas as to size and price. Nevertheless, you reach an agreement.

The next surprise comes with the announcement of the book in a flier. From the whole volume, someone has picked the figure that is least typical of the contents. You can avoid these unpleasant surprises by being proactive: write the first draft of the blurb yourself. If the publisher uses a graphic on the book cover, make a couple of suggestions for their designer to consider.

When the proceedings appear, you discover (*a*) a spelling error on the back of the book; and (*b*) in some figures 'bleeding' letters and numbers. You had previously asked about it and were assured this would not happen.

Months later, you receive a list of journals which received copies for review. It contains journals you did not recommend because they do not review books (and, of course, don't return the copies); and the list also contains journals whose readers would hardly have any interest in your book. But the most important journal whose editor has been waiting for the book has been omitted. As a result the most important review will be two years late, and recent, semi-competitive publications are already on the market. Thus the sales of the proceedings will never be what they could have been. To get reviews published, you need to work with the marketing department of the publisher and help to identify the specialist journals. You can send letters to the reviews editors telling them to look out for your book, and you may want to suggest reviewers.

In the preceding paragraphs, I have reported experiences with private enterprises. What about publishing with a government press? My advice is: forget it, unless your situation is special. This could be, for example, if your proceedings will appear as a numbered report of a US government publication series. As a general rule, however, recall that you will be dealing with a bureaucracy; and that bureaucrats develop an uncanny skill for dodging responsibility. In other words, if things do not work out with a government press, complaints will get you nowhere, contract or not. So, why take chances?

What about a university press? There are university presses that are run like private business; some of them, like the publisher of this book, do an excellent job. However, there are also others, often run by non-private universities or museums. They can delay your publication for years. If you think this is exaggerated, go to your library, pull out some proceedings published by universities, and compare the dates of the meetings with the proceeding publication dates; but don't be fooled: not long ago, I received proceedings that were backdated three years. Before you consider negotiating with a university-owned press, check the most recent references quoted in papers with the date of their publications.

The most scurrilous situation I know of occurred when the almighty editor of a university press decided to mount the authors – a behavior by which the males of some mammals confirm their superiority over other males. Of course, he did it intellectually; by changing the zoological taxonomy, and inflicting his favorite words on other people's prose. And that infuriated the authors because his animal kingdom just wasn't theirs; nor did they feel comfortable with an excess of 'inasmuches' and 'moreovers.' As a result, the proofs were so full of authors' corrections that the printers could hardly find their way through the mess; and in some cases, they didn't. It is unlikely that the man would have lasted in a private outfit.

In the final analysis, what is important in the selection of a publisher? Probably, the following:

(1) The track record, as shown by fast publication and wide distribution of his books. For journals, you can get specific figures if you look up the annual 'Journal Citation Reports' of the Institute for Scientific Information, Inc. (Philadelphia, PA). They are derived from the combined 'Science Citation Index' and 'Social Sciences Citation Index.' They should be available in any major university library.

Of specific interest should be three sections: (a) the ranking of journals by times cited in a given year; (b) the 'immediacy index,' a measure of how quickly the 'average article' in a particular journal is cited; (c) the 'impact factor', a measure of the frequency with which the 'average article' in a journal has been cited in a particular year. An immediacy index of more than 0.300, and an impact factor of more than 2.000 are often considered good. Libraries use the 'Journal Citation Reports' as a guide for their acquisitions, and so do an increasing number of our colleagues when they select journals for their publications. However, since both immediacy index and impact factor vary with the scientific discipline and the scope of a journal, it is very important to refer to the subject category listings as well as the editor's comments on their use.

(2) The quality of photographic reproductions, if you anticipate contributions with low-contrast photographs. For these, the 'grid' (dots per area) will be important. Ask the editor of a journal with excellent microphotographs what

grid they use; then, see if the publisher you envision for your proceedings can match it.

(3)   The attitude of the editorial staff in charge of your project. Editors may change, and not all of them are created equal.

(4)   Last but not least, the experience of those of your colleagues who have recently dealt with the publisher.

8.3.6 Selection of contributors

As with coeditors, one cannot be cautious enough in the selection of contributors to proceedings. Sometimes, this creates a dilemma if one wishes to invite a person who is known as an interesting speaker, but also as a manuscript delinquent. Since hope for a miracle is not indicated, replacement of the person with one of his coworkers may be an alternative (see Section 9.3). If he is greedy, you can bait him with money: no perfect manuscript on time, no travel support (see Section 8.3.2).

If you do not know a prospective contributor well, make some inquiries before you invite him. After all, funny and not-so-funny things can happen between your first phone call to him, and the delivery of the final version of his manuscript. Much of the following story is not exaggerated although the names are fictitious.

You are organizing the Plenary Lectures for the 'International Conference on Interactions between the Sexes.' Naturally, you are appalled by the continuing domination of this world by older males, and you want your colleagues to learn something about the roots of this atavistic indignity. And so, you remember Dr Barker, that outstanding speaker. Oh yes, his talk at the California Association of Effective Psychiatrists (nicknamed the nutcrackers) was a smashing success. Fabulous, how he explained complex interactions; who would have thought that curbing of dogs by middle class males betrays hidden insecurity?

Unfortunately, it turns out that Barker is already booked for other meetings at the time of your conference. However, he warmly recommends his friend, Dr C. H. Anger, a most unusual senior scientist: rumor has it that he does actual laboratory research.

You know that Dr Barker would suggest only the best, and that makes a second opinion unnecessary. You call Dr Anger right away. A delightful telephone conversation reveals his imaginative, flexible mind, and he agrees to give a lecture entitled

'The Evolution of Male Sexual Strategies: The Tricks of a Trade'.

And now, let's look at the subsequent events.

January 5, 1988   You call Dr Anger to confirm his honorarium. During the conversation, he suggests that the title of his talk should be slightly modified:

'Male Reproductive Behavior in Vertebrates,' by C. H. Anger.

April 20, 1988   Telephone conversation with Dr Anger to confirm the title of his talk for the preliminary program. He suggests:

'The Role of the Testes in Vertebrate Reproduction,' by C. H. Anger.

October 5, 1988   The abstract of his talk arrives five days after the deadline, with a new title:

'Structure and Function of the Mammalian Testis,' by C. H. Anger, D. R. Opper, E. M. Bar-Assed, P. A. P. Erlover, N. O. Good, and A. L. Wayson.

The names read funnily without punctuation, but this is none of your concern. You must make sure that the abstract goes to the printer on time, and that it fits the camera-ready format. Right away, you note several typos; also, the coauthors' names and affiliations have left little room for the text, and so it extends beyond the allotted space. To save time, you have it retyped with a smaller letter size. You pray that the letters will not bleed when it is reduced by the printer.

October 25, 1988   A revised version of the abstract arrives by express mail (it is now 25 days past the deadline), and you hardly recognize the title:

'Structure and Function of the Rat Testis,' by C. H. Anger, P. A. P. Erlover, N. O. Good, and A. L. Wayson.

You wonder why two coauthors have withdrawn their names. At least, no retyping is necessary. You call the printer (at your expense), and you get a free lecture on timeliness; then you rush the revised abstract to the angry man (you pay for special delivery) and ask him to insert it instead of the earlier version. Then, you pray that this intervention will not mess up the abstract volume.

February 1, 1989   Dr Anger gives his talk at the conference. In his introductory remarks, he points out that because of the limited time available, he will emphasize structural aspects. Hence, a better title for his presentation would have been:

'Structure of the Rat Testis.'

February 5, 1989   The last hours of the conference. For days, colleagues have asked you *ad nauseum* why you invited him. Ever since his talk, you have avoided Dr Anger like a bogey. Finally, you run out of luck, and he hands you the manuscript, the one you hoped you would never see: 'I revised it carefully because you insist on a perfect version for the camera-ready process. And besides, you would never pay the honorarium without the manuscript. May I have my check now?'

    You hand him the check, take the manuscript and run to a hidden corner. Fortunately, an armchair is not far away. The text is several pages too long, the photographs are of poor quality, and handwritten corrections stand out. That means, *you* must have it retyped. And then you read the title:

'Some Aspects of the Ultrastructure of the Normal Testis of Healthy Male Rats,' by C. H. Anger, P. A. P. Erlover, U. S. E. Less, N. O. Good and A. L. Wayson.

April 20, 1989 The manuscripts for the proceedings have been reviewed, everything has been carefully edited, the index has been compiled. Thanks to your insistence, even Anger's manuscript has been accepted. Of course, you streamlined the title without asking the fellow. As you are about to send the volume to the publisher, you receive an express airmail delivery. With trembling hands, you open the envelope, and read:

'A Structural Approach to the Cytoskeleton of Presumptive Interstitial Cells of the Testis in Male Baby Doomsday Rats,' by C. H. Anger, P. A. P. Erlover, U. S. E. Less, N. O. Good, A. L. Wayson and R. E. Viser.

Slowly, you stumble to your desk and write a postcard, to be sent by sailboat:

Dear Dr. Changer:
   We herewith refuse the latest edition of your paper.
   Sincerely,
     For the editors: S. H. O. Veit

8.4 Session reports and printed discussions

Session reports ('*Tagungsberichte*') are an old tradition of some societies. Usually, they appear in the journals of these societies, and they can give a lively picture of what was said and going on. Decades later, they often make fascinating reading because they give insights into the way things were; and how and what people thought long ago ('*Zeitgeist*'). Depending on the person who took the notes, some are truly humorous, others just plain dull. Of course, the writing of these reports requires considerable tact and skill, and not rarely idealism. Today, the use of a tape recorder has made the job much easier, and one wonders how long it will be until some societies videotape their sessions. Perhaps, the greatest problem with session reports is that they are not *verbatim* reiterations. This invites minsinterpretations of complex scientific contexts, and trouble with those who do not share the note taker's sense of humor.

  Printed discussions, on the other hand, can be copied *verbatim* from tapes. In review volumes or proceedings, they are usually printed after the paper to which they pertain. For the sake of accuracy, they are sometimes not edited at all. This is resented by people who hate meaningless flattery ('I must congratulate you on your magnificent presentation, and I think nobody would have done a better job. It was a true privilege to ...'). Of course, one can ask discussants to approve streamlined versions of their comments, or to write down what they actually meant to say. But then, you may run into a new problem: questions and pertinent answers that no longer match. Which shows that our memory plays nasty tricks when we try to be smart in retrospective.

So should you have printed discussions of symposium lectures? Perhaps, if you can afford them financially (i.e., spacewise), and if you are not afraid of all the added burden that they may become. How about this answer for a lesson in ambiguity?

9

Selection of participants: how to lose old friends and make new enemies

9.1 General audience

Sometimes, an organizer must decide if a meeting should be open to all those interested, or restricted to a certain audience. Since restrictions often create animosity, it is better to avoid them if possible.

Meetings with restricted audiences, such as the Gordon Conferences, may be desirable for confidentiality and/or in-depth discussions. To keep bad feelings to a minimum, however, one should be honest about the purpose of 'closed' meetings, and the rules that apply. Blackball schemes have a tendency to backfire, sometimes to the point of ostracism of the organizer. As they say, every dog will have his day.

Of course, the difference between 'open' and 'closed' meetings can be blurred. For instance, when a society raises the convention fee for non-members to prohibitively high levels; or, when someone convenes a conference in a place with limited overnight accommodation, and fills the available rooms with friends.

The weirdest proposal I ever came across requested funds for a special conference. The audience was limited to the number of seats in a small room. And how were the participants to be selected? By inviting everybody interested to submit an application that was to be reviewed by the organizer – who himself had no research funds whatsoever!

Question: Why should anyone pretend a *de facto* closed meeting is 'open?' Answer: In general, public money is easier to get for 'open' meetings.

In some fields of research, growing numbers of participants have changed the family atmosphere of the conferences. Since this may be resented by the traditional audience, a compromise is called for, perhaps as follows: (*a*) divide the program into specialized symposia; held in parallel if necessary; (*b*) arrange minor social events (e.g., lunches, dinners or receptions) for special interest groups; and (*c*) hold general sessions and major social events for all participants.

If the number of participants of an 'open' meeting must be limited, the organizer usually has two options: (1) only persons who submit their full fees by a *deadline* are admitted; (2) participants are admitted on a '*first come, first served*' basis, provided they submit full fees with the application forms. In either case, some people will probably send the application forms with the promise that (*a*) their fees are on the way (their bank or university is so slow), or (*b*) everything will be paid upon arrival because their country has restrictions on payments in foreign currencies. This puts the organizer in a very awkward position. If he allows this kind of 'reservation,' he is being unfair to applicants who submit their fees but cannot be admitted because all spaces have been filled.

9.2 Participants in key roles

The term 'key role' is applied here to participants who are essential for the program. These include speakers, colloquium moderators, colloquium participants, workshop leaders, introducers of special lectures, session chairpersons.

9.2.1 Speakers

In the selection of plenary and state-of-the-art lecturers, first priority must be given to the *quality* of the presentation. Especially at international meetings, it would be unfair to expose the participants, often after a long and expensive journey, to poor lectures. No doubt, it is difficult to set up a high quality program that is truly balanced, both scientifically and nation-wise; however, nobody will gain if the quality of a meeting suffers because of political considerations. Organizers of major international meetings are routinely pressured to have a 'fair' representation of this or that nation, or group of nations, in key roles. When this means speakers of inferior quality, the leadership qualities of the organizer are taxed.

It is better to bow out of any kind of responsibility if a meeting is held in a country where non-scientific considerations shape the program. If they ask you to assist in the preparations of a conference, give the best advice and all the help you can give on an informal basis.

Sometimes, you wish to invite an excellent scientist whose English or pronunciation is bad. This does not create much of a problem if the person can read a well-prepared English manuscript; or, when a colleague reads the manuscript for him. In either case, however, it is important that the author participates, via a colleague who translates, in the discussion.

As pointed out in Section 3.1.1, Plenary or Main Lectures should be given by established scientists who can present a sovereign overview of a field; while, on the other hand, State-of-the-Art Lectures may be more appropriate for younger scientists who report on the cutting edge of research.

Usually, the choice of presenters of Short Communications is less critical; depending on the meeting, all participants may be eligible to make one or more such presentations. Special graduate student papers have been mentioned in Section

3.1.1.4, and problems created by conveying to Short Communications a status superior to Poster presentations have been considered in Section 3.1.2.

The quality and desirability of After-Dinner Talks and Closing Lectures have been discussed in Section 3.1.1.

Depending on the occasion, Plenary Lectures may have to be introduced. These introductions should not exceed ten minutes, nor turn into lectures of their own. Good introducers have: (*a*) knowledge of the field which is to be considered; (*b*) information about the person to be introduced; (*c*) good humor; (*d*) enough discipline not to go overtime; (*e*) the decency not to demand special financial support. Ideally, the introducers are colleagues who are primarily involved in other events of the meeting.

Occasionally, one would like to invite a speaker even though it is doubtful whether he will be able to show up because of political or other uncertainties. In this case, see if he can team up with a colleague who, as a coauthor, will give the talk if necessary. It is less desirable to have the manuscript read by someone who is not able to answer questions during the discussion period.

9.2.2 Colloquium moderators and panelists

The job of a *colloquium moderator* requires considerable talent. Above all, he or she must exude authority, since one of the main jobs is to prevent talkative discussants from going overtime. Some people are born moderators, some can learn the role; others will never function under the dual pressure from panelists and audience. There is no sense in persuading a shy or timid person to act as moderator; nor will it help to pick someone who would use the opportunity for an ego trip. You want a humorous person with tact, good perception of the mood of both panelists and audience, and a thorough background on the issues to be discussed.

Good colloquium *panelists* are not easy to come by, either. The panelists should function as members of a team that jointly strives for success. Some people will never grasp the idea and confuse their role with that of a speaker, pope *ex cathedra*, or clown. And of course, the opposite is not uncommon: the person who hardly dares to speak up though he could make a major contribution to the discussion. If his problem stems from a repressive academic system, a tactful private 'pep talk' before the colloquium may help. At any rate, it is advisable to make a background check before inviting the panelists. Also, one of the panelists should be qualified and willing to substitute if the moderator drops out for unforeseen reasons.

9.2.3 Workshop leaders and participants

In general, it is easier to lead a Workshop than moderate a Colloquium. The informal atmosphere makes it more enticing to ask questions and change one's opinion.

Younger scientists may benefit from serving as a *workshop leader* before they try to moderate a Colloquium or Forum. To be on the safe side, there should be at least two workshop leaders. The more experienced one should open the workshop and announce the rules. The other one may take over once things are running. Understandably, the leaders of Socratic Workshops should: (*a*) have a good measure of humor; (*b*) be able to bring the evening to a harmonious conclusion. That is not a job for a sourpuss. Further aspects have been discussed in Section 3.1.4.3.

9.2.4 Forum moderators

The *forum moderator* has to deal with a crowd that may include the public. Also as pointed out in Section 3.3, there may be major difficulties. Hence, moderating a Forum takes experience, eloquence, tact and a good perception of mass psychology. A thorough preparation is called for; this means familiarity with the facts, ready answers to questions that are likely to come up, and a flexible strategy that prevents boredom as well as serious confrontations. These are good reasons to choose the moderator as early as possible.

The discussion in a Forum may not benefit from two comoderators, especially when controversies are to be expected. In analogy you can't have two commanders-in-chief directing an army in battle; but you need a second in command, ready to jump in when the general becomes incapacitated. It may be useful to have a colleague as a stand-by in case the moderator drops out at the last moment, or botches his job royally. To make the role of the 'spare' more attractive, he may also be called 'moderator,' with the understanding that, if all goes well, he will be in charge only of the opening remarks and the introduction of the experts (if such are invited).

9.2.5 Session chairs

Sessions should be chaired by two *session chairpersons* of equal standing. These chairpersons must be good time keepers, a very critical role in parallel sessions with Short Communications (see Section 3.1.1.4). This demands selection of people who can: (*a*) politely but firmly stop a speaker, if necessary; (*b*) decline any discussion if the speaker has used up all the time for the talk; (*c*) refuse permission when discussants wish to show slides of their own (something that should be forbidden and announced accordingly in the program, anyway); and (*d*) shut up pompous people who are famous for unnecessary and irrelevant comments. Obviously, chairing Short Communications is not recommended for shy individuals. At any rate, the job of the chairpersons will be easier if there are audiovisual devices that alert the speakers to the time available.

Chairing lectures of longer duration is usually less difficult, and it can be used to familiarize younger scientists with the art; besides, the honor may help them in their quest for travel support.

However, the prospective chairpersons must be informed as early as possible. Here is one example of how not to do it: I had organized a small symposium for a large international meeting, a job that involved considerable time and resources for planning, correspondence and fund raising. Upon receiving the program at the registration desk, I found myself teamed up with a cochairman who had not been involved in the preparations at all. Surprise! Fortunately, he happened to be a friend of mine.

9.3 Whom not to invite

If persons are known to (*a*) cancel their talks at the last moment, (*b*) fail to hand in manuscripts on time, (*c*) create unpleasant situations, or (*d*) try to cheat organizers and/or hotels, you don't invite them, and most certainly not as speakers. We are talking here about a very small minority whose actions, nevertheless, can seriously affect your mental health, the success and congenial atmosphere of your meeting, and your budget. If you are lucky, confidential talks with organizers of preceding meetings will help you to eliminate certain people from your program. Let's look at some prototypes who you don't need at your meeting.

The *megalomaniac* is usually a person, often a male, who cannot handle recognition. When you invite him to give a talk, he will probably promise to think about it. At any rate, because of his expenses for prior commitments (imaginary as they may be), he insists that you pay his travel and other expenses in advance. If you agree, he will let you dangle for a while and eventually accept.

And then, your real troubles start. His abstract is late. At the meeting, he delivers his talk and disappears. Why should he participate in scheduled discussions, give younger colleagues a chance to ask him questions? Needless to mention, he comes without the promised manuscript; and the chances are he will never submit it.

You may be much better off inviting one of his good coworkers. This person will probably know more about the work because he or she works at the bench. Perhaps, he or she is also looking for a position elsewhere. For people in search of a job, a Main Lecture is an ideal opportunity to advertise their qualifications.

Beware of the notorious *last-minute no show*. Certain people accept invitations but routinely cancel their talks too late for a change in the program – if they cancel them at all. Some of these people seem to panic as the time approaches to face their peers from the stage. The opposite extreme is the person whose ambitions have reached the border of paranoia. This person will only attend a meeting when certain to meet influential powerbrokers. When confronted for the first time with such a case, I did not heed the warning. As a result, the fellow cancelled his participation three days before an international meeting, and there I was, with an empty slot for a main lecture, and praying that he would return his travel support.

Unfortunately, the *habitual manuscript delinquent* is a rather common species. Normal human beings can be persuaded, or shamed into submitting their manuscript within a reasonable time. Habitual manuscript delinquents, on the other hand, are totally oblivious to the aggravation and loss of time and money they cause. There are cases in which they have delayed publications for a year or more.

And then there is the *travel fund bargain artist.* I have experienced this type in three varieties. The first one announces at the registration desk to everybody who cares to listen (and many people do): 'I was promised that my registration fees and room were to be paid for by the organizers. Now I am here . . .' Variation No. 2 insists that he can only pay with a personal check from a certain bank in a certain other country. This, of course, despite the announcement in the preliminary program that only one currency will be accepted.

What can you do with No. 1? Embarrassed, you take him/her aside and plead: 'We obviously have a misunderstanding. I am sorry. Please, use your credit card.' 'My credit card is not honored here!' 'What if we do not charge you registration fees until you are back home?' Of course, now you are lying to yourself since you just waived the fees. But your suggestion is not satisfactory anyway, and so you advise your friend to borrow money from colleagues. And off you run; sick to your stomach.

How about case No. 2? He creates delays at the registration desk so that your staff call you for help (a long line of participants, tired from travel, waits behind him). In desperation, you accept his doubtful check for the registration fees, and you are glad that your friend settles with the hotel, somehow, without getting you involved. Months later, the check will pop up again: as you close the account of your meeting, you discover that the bank was unable to collect the money, but charged you generously for their trouble.

The third type of travel fund bargain artist is truly disgusting but fortunately rare! He professes abject poverty and everything else that makes you cry, and moves heaven and hell until you finally give him travel support. This you do with mixed emotions, knowing that you are subsidizing the not-so-great investigator with other people's fees. Then, at registration time, your friend appears with his happy wife who is looking forward to a nice shopping spree.

There will hardly be a meeting at which people will not try to get out of payment. A common example are *cheats* trying to attend sessions, or social events without paying registration fees. The obvious remedy is to have guards who check name tags and rigorously refuse admission. However, some discretion is necessary: an honest person who forgot or lost his name tag, or a participant arriving after the usual registration time, should not be embarrassed. Things get more complex when people 'borrow' name tags from other participants. A typical dodge is for someone to enter the meeting area and send his own name tag, via an accomplice, to a friend waiting outside. I have been to conventions where this was epidemic and must have cost the organizers thousands of dollars in lost revenue. The organizer cannot ignore this situation if honest participants insist that he interfere. This can become your most hated memory, especially when you must deal with well-known individuals at a small conference.

It is unfortunately true that some of our scientific colleagues will try tricks at meeting after meeting, though not necessarily the same ones. I was once forewarned by a colleague that a person had left a previous conference with an unpaid dormitory bill. When all her time-consuming correspondence did not make me pay her travel expenses for another conference, she still appeared and involved me in a friendly conversation. Thereafter, she went to the registration desk and declared that I had waived her conference fees, etc. My staff were surprised. However, they were pressed for time because of the long line of waiting people and could not contact me immediately. The result was that the cheat got all the registration material, including tickets, program and abstracts, for free. Overburdened by other tasks, I admitted defeat when I found out, but prevented further losses: I informed the hotel which made her pay her bills daily.

10

Committees: you have to live with them

10.1 Mini-Machiavellian management

10.1.1 Making and breaking of committees

According to an old adage, the camel is a horse designed by a committee. It may be difficult to express better the feelings of many people who have had to deal with committees.

Before you consider working with committees recall that, in general, they serve one of four functions: (1) to come up with something useful; (2) never to produce anything of consequence; (3) to fulfill a requirement without making waves; (4) to hide foregone conclusions behind a collection of yes-men.

The first type of committee is often set up, and may even deliver something meaningful. It requires qualified and cooperative members, and it functions best when chaired by an enlightened dictator.

The second type is useful when a problem requires benign neglect. The more members it has, the less likely it is to come up with something serious. Committees of the third type often exist in the form of editorial boards for the conference, or they advise on tantalizing matters such as ceremonies and protocol.

The last type of committee may better be termed 'pseudocommittees.' Usually, they are *ad hoc* collections of friends, or people who depend on the grace of the chair.

Someone experienced in dealing with committees will probably subscribe to the following rules:

(1) Never set up a committee unless and before it is necessary.
(2) Select committee members very carefully choosing persons who genuinely will participate and are qualified to do so.
(3) If a committee is meant to function, keep it as small as possible, but as large as necessary.
(4) If a committee does not function, dissolve it.

A leader in a critical position needs advice without being forced to abide by it. This is why the organizer of a meeting should consult informally with knowledgeable colleagues, well *before* he considers involving committees. Rejection of a suggestion may be easier on a personal basis than in a committee meeting.

In practice, this may lead to a small 'kitchen cabinet,' a system US Presidents have used for many years.

We can learn from the big guys; for example, when we must set up a 'political' program committee. In this case, important decisions should be made with experienced advisers, *before* the committee is established. The next job is to place qualified and cooperative people in all committee seats that can be filled without restrictions. Only then has the moment come to convene the committee and appraise them, diplomatically, of the *faits accomplis*.

What if there are disagreements that cannot be resolved during committee meetings? Under certain conditions, the following approach may help. Shelve the controversial issue as fast as possible and bring it up, very casually, at a social event. For instance, at your home after a pleasant dinner, while indulging in some good wines. Creatures tend to be more agreeable when they are guests in the lion's den. Start with some issues all will agree upon, and then sneak in the controversial topic.

The organizer cannot be careful enough in the selection of committee members. A single obnoxious person can turn each and every session into a hell that begets nothing but problems and delays. There are tell-tale signs that indicate whom not to pick for your committee, if it is meant to function. It certainly pays to avoid the following.

The bully   He has a hunger for power. Unless he wants to pull a trick, he comes ill-prepared to committee meetings; insists that he knows best; tries to make other people do his job; and gets very upset if things don't go his way. If things work out despite his opposition, he will take full credit for the success.

The abominable no-man   Some people say 'no' whenever something new or exciting is proposed. These persons seem to suffer from congenital lack of imagination and a fear of decisions. When given responsibilities, they create havoc by sheer inertia. On committees, they are like sand on a piston, and that is sometimes their assignment.

The procratinator   As the time of your event approaches, fast decisions must be made. This is when you don't need a committee member who routinely insists on more time to think the issue over, or that advice from elsewhere needs to be obtained. If you push for an instant decision, he will vote against you, or he may abstain, or leave the room in protest before the vote.

What do you do when you inherit a powerful, but unwieldy committee? For example, you can't develop a meeting program with an international advisory committee of 56 members, representing 28 nations. When faced with this situation I found, serendipitously, a fast, elegant solution: I asked each one for written suggestions on certain issues. The result was as follows: about half of the people never answered, and the opinions of the other half were split about 50 : 50 – a classic case of committee sclerosis.

So, I thanked them for their cooperation and set up a manageable kitchen cabinet, consisting of six persons. Of course, I kept the big committee informed and drew on its members' individual expertise whenever necessary, but I did not drown in a deluge of international mail.

From the preceding, it would appear that the organizer of a meeting should work with an optimal minimum of committees; and must not confuse staffing with stuffing when appointing committees.

What committees are actually needed? Sometimes, none. If you run a minor conference, you and one or two friends may be able to handle all there is to do, including applications for travel support, and other fund raising activities.

However, the larger the meeting, and the more complex and varied the expected scientific contributions, the greater the need for some kind of formal committee.

10.1.2 Mundus vult decipi

The above quote from a pope early in his career suggests that people want to be fooled. The organizer should benefit from this Latin tag during fund raising, and when he must project authority. This brings up the question of names, ranks and titles. As we all know, our fellow men and women respect symbols of power and honor. This demands impressive names for committees. Words like 'national' or 'international' are particularly effective. You may be more successful speaking for the 'International Committee on Atmospheric Conditions' than the representative of the 'Brotherhood of Weathermen.'

In the same vein, you may be more effective as the 'President' rather than the 'Organizer' of an International Symposium. But don't overdo it; the majestic is close to the ridiculous. Don't call yourself 'CEO of the Culinary Institute' when all you own is a spoon and a fork. Also, you would not want to style yourself 'President' of a small regional conference. 'Convener' or 'Organizer' will definitely do.

On the other hand, be generous when it helps your colleagues in obtaining travel funds. Don't hesitate to splash titles like chairperson, moderator, convener, organizer, panelist, invited speaker, invited presenter, or whatever comes to your fertile mind, as long as this does not entail financial commitments for you nor deliberately misleads funding agencies.

10.2 Standard committees

10.2.1 Organizing and program committees

Assuming that you have decided to be the president of a major meeting, how many 'major' committees will you really need? Often, one committee will suffice if you hold the meeting at the place where you are located. In this case, the 'organizing committee' will be identical with the 'program committee,' and it will also function as the 'local committee.'

A single combined organizing/program committee may also suffice if a meeting is regularly held at the same convention hotel with an experienced and reliable staff, or when a society has a permanent staff that takes care of all technical details. In this case, no 'local committee' may be needed. However, a good rapport between the staff of the society and the committee is a must. This requires, from the beginning, a clear understanding of the respective roles and duties.

When a society has different divisions that develop their own scientific programs, the role of the organizing committee may be reduced to that of a coordinating body, and the president of the meeting may be hardly more than a figure head.

Remember that committees must be kept to an effective size. I look at an organizing committee consisting of more than six members with apprehension. For the program committee, three members may be enough, provided you can draw on the expertise of other colleagues when necessary. If you set up a large committee, you may soon be fighting with members who try to get friends in program spots where they don't belong. Conserve your strength; as the organizer you will have to deal continuously with people trying to meddle in your affairs. This can become especially painful if a society has a council or other advisory body.

An important factor in the choice of committee members is geography. If it is possible, select only members who live within driving distance from each other or a convenient place like a laboratory or home. If you must consult via telecommunications, use fax or the internet, particularly during the final, hectic phase of preparations. At this point, for example, nobody would gain from an international committee that could not function. Hence, it must be understood that, from a certain moment on, the organizer will have the authority to make all decisions that cannot be delayed.

10.2.2 Local committees

When a larger meeting is to be held at a place which the organizing committee is not familiar with, it may be necessary to set up a local committee. The charges of the local committee usually include advice on the venue, provision of equipment, recruiting of local assistants and volunteers (especially projectionists and trouble shooters), selection of recommended restaurants, and arrangements for receptions, excursions and other social events. One very important function of the local committee involves the housing of the meeting participants. As pointed out in Section 6.1.1, it is dangerous to rely on an office of tourism when choosing accommodation. The local committee should always ensure, by personal inspection, that the selected accommodation is: (1) close to the meeting place; (2) clean and comfortable; (3) worth the money. Only then should they decide if it is worth involving the local office of tourism.

For the organizer of the meeting, it is a *sine qua non* to know which responsibilities can be delegated to the local committee. This is especially critical at international meetings. Often, local committees are tempted to go overboard to gain recognition from their national authorities, and international prestige for their country's science. As a consequence, the audience may have to suffer lengthy or inappropriate speeches by government officials, and the organizer will have a tough time keeping inferior speakers from the 'host nation' out of the program.

It may be useful if the organizer makes important local arrangements jointly with the future chairman of the local committee, *before* the committee is installed. Though this may infringe upon the traditional activities of local committees, it leaves important chores for them; for instance, the flawless running of the sessions during the meeting. For the organizer, the advantages of such preemptory activity are that: (*a*) he will know that things are under control as much as possible; (*b*) he will have a good personal rapport with the chairperson of the local committee; and (*c*) time and money for telecommunications may be saved.

The job of a chairperson of the local committee is often less than rewarding. He or she will have considerable responsibility and only limited authority. If something goes wrong, the local chairperson will be a convenient scapegoat. He will be blamed for equipment failures, bad weather, thefts, robberies, and poor public transportations, to name a few. In big societies, his name may soon be forgotten, provided all went well. Why, then, should anyone take the job? Well, perhaps because it is an excellent learning experience for someone who plans to run a meeting of their own (recall Section 2.1).

10.3 Other committees

Committees on satellite symposia As detailed in Chapter 19, satellite meetings must be carefully coordinated with the main meeting. Depending on the circumstances, the committee for a satellite meeting can be a subcommittee of the program committee; or, at least one member of the program committee should also be on the satellite committee as liaison.

Award committees Sometimes known as 'juries', these often have great responsibilities. If they are perceived as unfair, this can lead to internecine warfare. I know one case in which they strongly contributed to the virtual paralysis of a society.

Sometimes, the organizer of a meeting has little interaction with the award committee(s). On other occasions, he is expected to raise travel funds etc. for prospective awardees before he knows whether they come from nearby or another continent. Hence, he must insist on an early decision.

The organizer may also have to deal with committees that give awards based on presentations at his meeting, for example, for the 'best' poster presentation. Since he may be blamed if something goes wrong, he may have to ensure that the respective award committees do their job (especially when the chairperson of a committee is inexperienced). If a single competing talk or poster is not evaluated, the committee has failed. The committee will also fail if its members apply differing criteria.

'Standing committees of societies'   These do not have to be considered here as long as their charges do not directly affect the organization of the meeting. Otherwise, rule (4) of the following paragraph applies.

Advisory committees   Inherited 'advisory committees' or 'councils' can be a source of endless aggravation. In order to keep muddling and meddling by members of these committees to a minimum, the following rules should be agreed upon: (1) The organizer of the meeting (president, convener or program officer) has ultimate responsibility, and with it ultimate authority. (2) The 'advisory' committee serves as a resource of advice and assistance to the organizer. (3) From a certain point on, the members of the 'advisory' committee will communicate with the organizer only when asked by him, or when there is a truly important reason: no more 'suggestions' of speakers, or changes in the program. (4) If a society has independent or semi-independent divisions, or steering committees, their authority and share of the meeting time are determined *before* the organizer accepts his job.

11

Accompanists: you better love'm

If you wish to attract as many participants as possible, you need a meeting site that is appealing to accompanists. However, a large number of accompanists requires the design of a special program for them. This is not a job to be taken lightly. Accompanists come with certain expectations; if they are disappointed – and this can happen easily – their mood will affect their partners. And you don't want this to happen.

Today, accompanists will be of either sex, a fact that requires some consideration. For example, male accompanists will not be interested in a lengthy fashion show (for outer garments), or in places where women try on and out everything from rings to shoes, handbags, belts and muu-muu. On the other hand, a local baseball or soccer game may not attract too many female accompanists. Also, the organizer may have to bow to tradition; for example, when the annual convention of an older society has a 'hospitality suite' with complimentary beverages, cookies etc. If you deprive the regular accompanists of that cozy place for their small talk, you may as well be dead. So, bite the bullet and pay for the dark fluid that is often mislabelled 'coffee' as well as soft drinks.

Misconceptions can easily initiate a chain reaction of complaints. At a meeting in a major city, older accompanists were disappointed that the downtown(!) convention hotel was not close enough to the main shopping area. The distance was about seven blocks, and that was not considered 'walking distance.' From this experience, it appears important to point out the distance between meeting site and shopping areas in the early announcements.

Bored accompanists can be dangerous to the success of a meeting. Inevitably, they will try to induce your participants to skip sessions and join them elsewhere. If you don't like half-empty meeting rooms, you had better keep your accompanists happy. Probably, the best way to achieve this are half-day excursions that terminate after lunch at an attractive restaurant with affordable prices. This keeps the accompanists entertained, and out of the hair of your participants. After such excursions, most accompanists will enjoy some rest during the afternoon.

Set the departure time for the accompanists' trips so that it is fifteen minutes *before* the start of the morning sessions, and you have solved a major problem:

your members will be on time for the first talk. Next, consider your options and make the first excursion not too demanding; include an introduction to local shopping opportunities, visits to the market and/or bazaar (provided they offer things worth seeing or buying), a small museum or garden, a winery or brewery, and perhaps one or more interesting buildings. On the following days, more demanding trips can be scheduled. Always try to offer a variety of events; it helps to combine, for example, culture, local craft and nature. However, if you have scheduled a joint excursion for both participants and accompanists, make sure that it will be the highlight for all. If you have foreign visitors, they may wish to have one morning or afternoon for a shopping trip on their own (usually towards the end of the meeting). Should this be dangerous (think of pickpockets, holdups), you have to provide guides.

Whenever possible, add something to the program that makes people feel special. Something that is not usually shown to the public, such as a collection of ancient artifacts, rare fossils, a stone age cave, a freshly excavated historical site, a mint producing gold coins, the interior of an ultramodern battleship, an astronomical observatory; or private property, from a game park to an art collection.

Children are a special type of accompanist. If the venue is not suitable for them, this should be obvious from your announcements. For example, an isolated medieval castle with limited bathroom facilities, and neither a sandbox nor a swimming pool is not an ideal place for children; especially, when the participants of the meeting enjoy noisy parties with hard liquor that last into the wee hours.

On the other hand, if you schedule a meeting during school vacations and anticipate the attendance of many children, this must be considered in the choice of lodging, and in the design of the program. As a rule, it is advisable to lodge parents with children in larger hotels that offer a variety of entertainment regardless of the weather.

A large swimming pool is almost a must if you expect accompanying children unless, of course, there is a sunny and safe beach with lifeguards (and no drug dealers). If you are lucky, the hotel may organize water ball games and similar amusement.

Provisions for indoor entertainment should include: several TV sets that are located so that different channels can be watched; games that can be played by two or more children; and video games.

For more diversion, see if you can organize, or recommend visits to an aquapark, safari or other fun park, an ice rink, a zoo, an alligator farm (with wrestling performances), a major aquarium or 'marineland' with shows, a planetarium or a technical museum (with devices that can be operated by visitors). In some places, hotels will sponsor crab races, or a community will hold frog jumping contests. Older children may enjoy excursions on horseback, nature walks, shell collecting, airboat rides (Florida Everglades), snorkeling excursions to a safe(!) reef, or a dive in a submarine built for sight seeing. Last but not least, speed boat rentals, water skiing, wind surfing, scuba lessons, bungee jumping and a dancing club may greatly contribute to the happiness of teenagers, provided the prices are reasonable, or the parents ready to pay anything for a few undisturbed hours.

For the organizer of a meeting, children are not necessarily an undiluted pleasure. The most common problems stem from the belief of parents that children may participate in all kinds of events for free. A typical situation arises when buses are chartered for the exact number of persons who paid for seats. All of a sudden, the bus driver refuses admission to persons who have valid tickets and you, the organizer, are called to intervene before a heated debate gets out of hand. You count the persons and discover that several children occupy seats for which their parents did not pay. The situation is clear, but how do you handle it? You want to save the face of the parents, and yet keep the rightful owners of the seats happy. In the case of smaller children, the obvious solution would be the lap of their parents; alas, you are dealing with teenagers, and the bus trip will take hours. Caught in this dilemma, I was ready to give my seat on the bus to one bumpee, and invite the others to come with me in a rented car (and, of course, refund their tickets). At the last moment though we discovered some empty seats in another bus.

A similarly unpleasant situation arises when parents argue with security staff who refuse to admit children without tickets to a social event. Usually, the parents use one of the following arguments: (1) the children left their name tags in the hotel room; (2) there is no sense in paying for children who will only participate in this event. In this case, there is little you can do but play the gracious host — and wonder what the parents who did pay for their offspring will say when they find out.

12

Office and staff: don't take chances

There is one overriding principle for the selection of your office staff: *a few thinking people*. Do not fall to the temptations of status display and hire people you don't need; and don't try to save money by hiring cheap labor. You will be better off paying good people overtime than employing helpers who need all the help they can get.

The ideal person for the office of a smaller meeting would be a secretary who has organizational talents, writes flawless English, is familiar with scientific terminology, is a good proof-reader, is experienced in the use of computers; and, above all, is reliable. Unfortunately, they don't always make them that way.

For a larger meeting, you may split the work between an assistant and a person mainly involved in typing. The job of the assistant includes the mailing of announcements and various types of forms, book- and budget-keeping, monitoring the timely submission of payments and scientific material (e.g., abstracts, questionnaires), and answering routine letters and e-mail. Ultimately, the assistant will also be in charge of the registration desk, even if it is staffed by employees of a professional service, or of a society (see below). The typist will handle most of the typing, from letters to forms, abstracts, manuscripts, etc. Familiarity with word processing is essential, particularly since the typist will have to update and correct continuously a list of addresses that can be transferred to mailing labels. The respective roles of the assistant and typist must be clearly understood from the beginning, and one of them must be replaced instantly if they cannot cooperate.

If you start your preparations early, and your meeting is relatively small (e.g., a regional conference), you may be able to manage with part-time labor. For example, a technical assistant or a junior scientist, and a good typist with a regular job, could handle meeting matters during evenings, and/or on weekends. However, meetings with several hundred participants and major international meetings require additional help. If you are lucky, you will have good graduate students who can be involved without it interfering too much with their research and course work. If you must use a secretarial service agency, this can be very expensive and frustrating. Usually, their staff are not familiar with scientific jargon, and they may become very defensive when retyping is repeatedly required. After a while, it is no longer funny when they routinely confuse ligation with litigation, planet with plant and organism with orgasm.

As the time of the meeting approaches, you will depend more and more on the experience and memory of your coworkers. No matter what computer you use, you cannot retrieve files that were never created. This is one reason why you should have reliable people working with you from the early stages of the preparations to the final closing of the budget, which may be long after the meeting (see Chapter 16). It is not smart to change horses in mid-stream.

A professional organizer or travel agency can be helpful, provided they are experienced and trustworthy, and you can afford them. Under ideal conditions, an executive officer of a society will assist you in all routine matters of the meeting so that you can concentrate on scientific aspects, fund raising, etc. However, the staff of a professional organizer, or of a society, should not make the final decision when certain people do not pay on time, not in full or not at all; or try to pull tricks at the registration desk (see Section 9.3). Also, the collection of manuscripts, if due at the registration desk, demands that either you personally, a coeditor or another 'enforcer' is present.

If you are unlucky with your hotel, angry participants may insist that you, their host, get into the act, even if all arrangements with the hotel have been made through a professional organizer. Typically, this happens when there are problems with rooms, rejected checks or credit cards.

If you run your meeting without the help of a professional organizer or the staff of a society, you will ultimately have three different crews:

(1) assistant and/or typist;
(2) the staff of the registration desk;
(3) technical staff.

The assistant and typist need to be available (though perhaps only part-time) from the early preparations until all the books of the meeting are closed. The assistant also needs to be present at the meeting, and readily available at the registration desk.

The staff of the registration desk must be present as long as the desk is open, and fluctuate in number according to the needs. Your people at the registration desk must be thoroughly familiarized, at least one day in advance, with their duties. These include not only handling of routine matters of the registration, but also advice to participants who wish to change money, or need information on local traffic, restaurants, cultural events and entertainment, and recreational facilities. All too often, participants get irritated when they cannot get clear answers from the meeting staff. You can never have enough maps and other information at the registration desk, especially when the employees of the hotel are poorly trained and/or unfriendly. At major meetings, don't forget a telephone directory that has instructions on international calls, and provide note pads plus pencils so that people can jot down the information.

Usually, graduate students or employees of a host institution can be drafted for the registration desk. If they volunteer, be generous and allow them to attend some sessions and social events, as long as enough persons remain at the desk while it is officially open. However, you must know with whom you are dealing. In some countries, students may not comprehend that volunteers must be punctual and reliable even though they are not paid.

The technical staff must be knowledgeable and reliable. At major meetings, they will probably cooperate with the professional staff of the meeting facility, and this may test your managerial skills if the latter are unionized, or just nasty to what they perceive as competition. Depending on the size of the meeting, your technical staff may consist of three different crews: (1) audiovisual staff (who also check flipcharts and blackboards, poster boards and supplies, and make sure that all speakers have clean drinking water); (2) security staff checking admission to meeting rooms, hired buses, etc.; (3) trouble shooters helping wherever needed, from emergency purchases (e.g., crayons, markers or pointers that disappeared overnight; push pins for posters) to projecting emergency messages, instructing taxi drivers, accompanying sick people to a physician.

13

The budget: a jungle with pitfalls

13.1 Initial consultations

When preparing the budget of a meeting, it helps to consult with those who have organized a similar event before. Ask them for a general, rough breakdown of their budget, and suggestions they may have for you. But don't insist on seeing their files; that may be the end of their cooperation. The organizers of a scientific meeting, like anyone carrying major responsibilities, may have to make pragmatic decisions. Just think how you, as a department chairman, would react if a visitor asked you to show him the files with individual salaries and budget allocations, or how you would feel if he were to ask you to mail him copies. The issue here is not irregularities or unfairness; the issue is that tough decisions, based on complex circumstances, can easily by misinterpreted or misrepresented.

13.2 Instant savings

One of the first steps in your budget preparation should be an estimate of the savings. As good old Benjamin Franklin noted: 'A penny saved is a penny earned.' The sad fact is, however, that many organizers inflate the meeting costs by neglecting opportunities to save money, and then are unhappy when people do not come because they can't afford it. How does one save money right from the start? First of all, one does not spend money on unnecessary frills such as memorabilia, expensive posters or announcements on multi-colored paper. Just think how many hours of typist's time you could pay for with the funds allocated for this self-glorification! If a photographer requests permission to take pictures, allow it with the proviso that he bears the financial risk. Do not commit any funds, unless you are ordering official group photographs.

Second, and more importantly, one avoids intermediaries, such as travel agents and professional organizers, whenever possible. These intermediaries are unnecessary if you have enough support and resources to handle the non-scientific part of the preparations. For example, if you are in charge of the annual convention of a major society, the executive director (or equivalent) of your society should be able to take care of most of the 'technical' details of the meeting.

If, on the other hand, you are a busy person, there is no executive director, and money is no object for your society, you can hire a professional organizer. This will save you time, and the logistics may be off your back. Probably, the professional organizer will explain that he actually saves you money because of his experience and connections. He may even claim that he can help you raise funds; and perhaps, he will guarantee a minor social event, free of charge. If, after a thorough background check, you have found the right person, this may all come true. But in the end, more likely than not, you will still try to raise funds through your own connections. And, of course, the congress hotel or the host city are often prepared to throw a free reception for the participants anyway.

A low-budget scientific society should decide on hiring a professional organizer after a comparison with the budget of the preceding meeting. The primary criterion is the registration fees. If they would have to go up substantially to maintain the standards the participants are used to, you can't afford to hire expensive assistance. Another criterion is the cost of excursions.

The latter tell you something about a prospective professional organizer. If he quotes you the same price for a bus seat as that listed by the local travel agencies, beware! On the other hand, a good professional organizer will be concerned about his reputation, and he will not accept a deal he cannot carry out to mutual satisfaction.

If you decide against a professional organizer, it may be tempting to ask a travel agency to handle housing, social events and excursions. If they oblige, you have unburdened yourself of a lot of work. Fine. But keep in mind that they make money from their services, and that their interests are not necessarily yours. Also, when they offer something special, there may be strings attached. Once, the travel division of a department store offered a free sight-seeing trip through their city, provided our accompanists would first spend an hour browsing in their store. Sounds fair? Wrong. If we had agreed, there would have been no time left to visit several historical sites of the town, which happened to close for the siesta.

There is hardly a business with more intermediaries than tourism and related activities. Greasing palms seems to be general practice, which means that you, and/ or your local committee, must be on the alert. If your budget is tight, get the prices of the local office of tourism and travel agencies; then, see if you can't get better deals by direct negotiations with hotels, restaurants and tour operators. This is not always easy, and in some places, dirty tricks are rife. As a rule, however, hotels, restaurants, and tour operators are more amenable to special deals if they don't have to pay travel agents. On two different occasions, I got a particularly good deal for bus excursions: about 30% less than the usual tourist fees, *without loss of services.*

In summary when dealing with the tourist industry, remember comrade Lenin who noted that 'confidence is good, but control is better.' He was an expert on organizing meetings, and he did not use travel agencies.

Registration fees can be kept down considerably if there are no charges for meeting rooms. Often, this can be arranged by holding the meeting on the campus of a university during weekends or vacation time; by arranging with the official meeting hotel for free rooms for sessions; or by special arrangements with congress centers, for example, when a congress center has to be heated during the off-season anyway.

On the other hand, don't invite a meeting to an institution that does not provide free meeting rooms, unless there are considerable trade-offs. If they are greedy, they don't deserve their name on your program. Also forget about a hotel that charges for meeting rooms; either they function as the official congress hotel (which means free meeting facilities and special room rates for participants), or they are not worth your consideration. If you are informed by the director of a government-run congress facility that you must (*a*) make a substantial downpayment three years in advance, (*b*) pay outrageous rental fees, and (*c*) pay for shuttle buses to the hotels (which are all beyond walking distance) – well, then let him rent the place to corporations that outdo each other in wasting the money of their shareholders.

An important means of reducing your participants' expenses is special room rates. However, hotels may demand a guaranteed minimum number of reservations, with deposits, by a certain date; so, make sure you have enough information before you enter into negotiations. During the pre- or post-season, when prices are low, the discount can be considerable. Naturally, your bargaining position is weak if you schedule a meeting for the peak season. Always keep in mind that you want the congress hotel to give you: (*a*) free meeting facilities; (*b*) low room rates for your participants; (*c*) complimentary rooms (one free room night per 50 paid room nights); and (*d*) most likely also specially-priced breakfasts and luncheons.

In many places, lodging in private homes ('bed-and-breakfast') is popular. It is usually cheaper than other accommodation, but you can only pray that your participants will be satisfied. The same holds for camping grounds.

Misunderstandings can happen when people are not familiar with the local customs. A colleague from the then East Germany attended a meeting in Japan. His authorities had prepared him, as usual, with hours of ideological hogwash about capitalism and imperialism, but told him nothing about the country. Once there, he checked into a family-run ryokan. He liked the little hotel very much, especially the wooden tub with hot water. He proceeded to bathe in it and cleansed himself with soap. Shortly thereafter, there was a big commotion which he did not comprehend. The heart-broken owners and the other guests stood before the empty tub whose expensive, clean water was used by everyone for a brief soaking – only after thoroughly freeing one's body from soap. Of course, the organizers of the meeting were called in for help – and graciously paid for the financial and emotional damage.

An important means of saving money is group travel at reduced rates. Usually, this applies to air fares. Find out what the airlines and travel agencies have to offer.

Do thorough comparisons. If a travel agency quotes you the regular price for a larger group (which has happened to me more than once), then they must (*a*) be stupid, or (*b*) think that you are stupid; or both (*a*) and (*b*) will apply. The official airline or travel agent of your meeting should give your participants a reasonable discount (provided you come up with a minimum number of passengers), and in addition a free seat for every 30–35 participants. Alternatively, see if your 'official' airline will give your participants a flat 35% discount when they present your certification that they are attending your conference. But beware: get things in writing. Let me emphasize once more that dealing with travel agents can be more tricky than trading horses.

Another possible source of savings may be the free use of projection equipment and/or poster boards, for example, when provided by the congress hotel or a host institution. Audiovisual suppliers and hotels charge sometimes exorbitant rental fees for projectors, screens, etc.

Avoid unnecessary payments for social events. See if a reception (e.g., a welcome party, or special reception for graduate students) or the coffee breaks can be sponsored by the congress hotel, a university, a city, a corporation or an exhibitor. But make sure that the event is going to be a good one. A welcome party with miserable drinks and cheap snacks can be devastating to the morale of the freshly arrived participants. However, be prepared to pay for the reception from the meeting budget, unless you can be absolutely sure of your sponsor's reliability, and the quality of the event (see below).

Portfolios and pens can be expensive when they are of good quality, and imprinted with your logo. Ask your 'official' airline, bank, institution, city, or some pharmaceutical firm to chip in for the privilege of having their name on the portfolio. For them, it should be good advertisement, but it may be difficult to convince them. You may be able to reduce the cost of brochures (program and/or abstract) by including advertisements.

If an international meeting is held in a country with an unstable currency, the organizers may decide to accept foreign payments only in US dollars or another 'hard' currency. This can become a source of considerable savings, but also of losses since the US dollar is continuously manipulated. At one meeting in a European country, I delayed all payments up to the last moment because the dollar went up from day to day. Nine years earlier, in the same country, I had run with thousands of dollars in cash from one place to another to pay bills (e.g., social events, excursions) before the drop of the dollar ruined the meeting budget.

Finally, in some instances you may be able to take advantage of tax exemptions. For example, if your meeting account is administered by a non-profit institution in the USA, you may not have to pay sales tax. This can amount to considerable savings. Assuming that the costs of proceedings have been included in the registration fees, consider the following example: special price per copy for participants: $80; total number of participants: 500; total payment for proceedings: $500 \times 80 =$

$40 000. Taxes due for regular sales: 5%; savings due to tax-exempt status of your university: $2000.

13.3 Expenses

In the following, the expenses for a scientific meeting are broken down into three components: (1) *Expected expenses*, which are reasonably predictable by category, though the exact amounts may yet surprise you. (2) *Contingency funds*, whose categories and amounts are only partly predictable. (3) *Cost overrun*, 'the safety net,' an important budget component. Experienced organizers may break down their budget into two components only, e.g., expected (projected) expenses and cost overrun; however, for the peace of mind of the less experienced, the pattern proposed here may be preferable.

The estimation of the expenses is difficult. In hindsight, many organizers of scientific meetings must have realized that their budget projections were utterly naive. I know two cases with a cost overrun of more than $10 000 for meetings of about 400–600 participants. In one of these, the organizer appealed to colleagues, including those who had not participated, to bail him out with donations. I doubt that this solved his problem. In another instance, the organizer asked scientific societies for help. Personally, I lost several hundred dollars after running my first international symposium (with about 100 participants); I had closed too early what appeared to be a balanced budget.

What is the usual mistake in the budget planning of scientific meetings? Under-estimation of contingency funds and cost overruns.

13.3.1 Expected expenses

The following listing summarizes the most predictable expense categories though some (e.g., 'honored guests' and 'customs') will not apply to all meetings. Depending on the particular circumstances of a meeting, additional items may be identifiable right from the beginning.

If a banquet is expensive, it should be handled as an independent budget item. Some people do not enjoy banquets, and others may hardly have enough funds to attend the meeting. These participants may object when the price of the banquet is included in their registration fees. Overall, the cost of the banquet should be kept low, but never so low that this compromises the quality of the food, beverages and entertainment. Remember that for many participants the cost of a banquet (and also of excursions) may considerably increase their initial payments.

When preparing the initial budget, you may select from the following list the items relevant to your meeting and insert the estimated amounts behind each one of them, add further items that come to your mind, and then calculate the total. When you do this the first time, you may be in for a surprise.

(1)   clerical supplies;
(2)   clerical assistance;
(3)   administrative assistance;
(4)   printing: announcements, program, abstracts, stickers, forms;
(5)   postage, including overseas airmail and express mail;
(6)   telecommunications (phone, fax, e-mail, cables, etc.);
(7)   travel expenses for certain committee members;
(8)   awards, prizes;
(9)   honored guests (from travel to banquet);
(10) travel support for participants in special or 'key' roles;
(11) equipment rental;
(12) special assistance: projectionists, trouble shooters, guards, travel guides; personnel for the registration desk;
(13) downpayments for reservations (rooms, buses, social events);
(14) final payments for the items in (13);
(15) bank charges;
(16) taxes and customs;
(17) portfolios, pens, writing pads, name tags and holders;
(18) coffee, tea and cookies;
(19) beverages (soft drinks) for speakers, session chairpersons, poster sessions, etc;
(20) shuttle buses;
(21) proceedings, if included in the registration fees.

13.3.2 Contingency funds

This component of your budget will include the less conspicuous and greatly varying expenses that may change your original projections before, during and after the meeting. Here are examples:

13.3.2.1 Before the meeting

(1)   Changes in prices for briefcases and pens, including work for the logo.
(2)   Changes in printing costs, especially of program and abstract volume (for example, when many more abstracts are submitted than anticipated).
(3)   Overtime pay for clerical and other work (don't count on savings with volunteers until they have done a *satisfactory* job; you may yet have to hire professionals).
(4)   Increases in the postage and/or shipping rates.
(5)   Telecommunications costs beyond the original projections, in connection with: fund raising; local arrangements at the meeting site; mail strikes; requests for

travel support; belated submission of abstracts; late dropouts from the program.

(6)   Fluctuations in the exchange rate of foreign currencies.

13.3.2.2 During the meeting

(1)   Petty cash expenses for: clerical supplies; poster board supplies (more hammers, pushpins, thumb tacks, tapes); minor items (projection equipment; spare frames for slides; foils and markers for overhead projectors).

(2)   Rental of additional equipment, or equipment replacement.

(3)   Payments for overtime, additional assistance for office staff and projectionists, or other supporting persons.

(4)   Taxi expenses for jobs that have to be done quickly (purchase of supplies; emergency rental of equipment; banking).

(5)   Funds for participants whose money and/or documentation were lost or stolen (especially when those concerned are foreigners). As a gracious and embarrassed host, you don't want to bill them later, for example, for the taxi to the police station and/or the consulate of their country; or for special long-distance telephone calls.

(6)   *Loans* to people: (*a*) who lose their credit cards, or have unacceptable credit cards; (*b*) who lose their money or cannot cash checks from foreign banks; (*c*) whose deposit for a room never arrived at the hotel. Of course, you don't know if they will ever send you the money once they are back in their homeland. Thus, the loans may end up in your category 'cost overrun.' I have had cases in which people did not even return the receipt for support they received.

(7)   Refunds for various reasons, (e.g., overpayment that was not noticed before the meeting).

(8)   New fluctuations in the exchange rate of foreign currencies.

13.3.2.3 After the meeting

(1)   Cost of telecommunications concerning late manuscripts.

(2)   Retyping (or mortising) of manuscripts that were supposedly camera-ready; if you are unlucky, this can mean retyping of all manuscripts submitted (see also Section 8.3.1).

(3)   Clerical and related expenses for answers by mail (in response to complaints, belated financial requests, etc.).

(4)   Clerical and/or administrative assistance in closing the books and account of the meeting.

(5)   Assistance in the preparation of the proceedings of the meeting.

(6)   Expenses not paid by participants. Classic example: bills for lodging in a dormitory without a front desk.

(7)   Fees for negotiating checks (especially from foreign banks).

(8)   Payment for proceedings to be mailed to sponsors (e.g., funding agencies), members of committees involved in the meeting, and other supporters (e.g., officials of the host institution).

(9)   Fees or payments for social events ('will be paid tomorrow') that were 'forgotten' by your participants; loans never paid back by participants; payments (registration fees) promised by their university or government that never arrived.

(10) Changes and differences in bank charges. During the past decade, even some European banks have become so 'efficient' that their charges are now high. Ask your bank how they leech incoming and outgoing checks, and make the expected losses a budget item. Otherwise, the closing of your meeting account may become an unforgettable experience. However, be aware that checks paid via different banks may be leeched in different ways, that is, some banks may charge more, and others less, than your bank.

(11) Fluctuations in the exchange rate of foreign currencies (the same item for the third time!).

This list gives some idea how seemingly 'minor' items can increase your budget drastically.

There is no general formula for the estimation of contingency funds. However, it is clear that their estimate should be higher for an international meeting. Also, the site of your meeting will have an impact. In a metropolis, almost everything will be more expensive than on a campus in the boondocks. On the other hand, emergency transportation from a small town to a hospital or airport could cause, at least temporarily, a major hole in your budget. With careful planning and experience, contingency funds should amount to 5–10% of the total budget.

13.3.3 Cost overrun

To the above, add a cost overrun of 10%. You may wonder why; after all, what can be so unpredictable? Here are examples from personal experience: The sponsor of a welcome party for 650 persons dropped out very late (for reasons not related to the meeting), leaving us with a projected deficit of $20 000. Another sponsor, expected to pay for the portfolios, also dropped out: add more than $2000 to the above. Then, problems with the postal service in one country delayed the printing of the program and abstracts; as a result, both had to be shipped by express air mail, overseas, to the meeting site: add more than another $2000. Then in the host country, we had to pay outrageous customs fees plus a fine.

You think that is enough? Wrong again! After the meeting, the publisher of the proceedings was bought by another company. This we noticed when they sent an invoice, suggesting that the shipment of the proceedings to the participants would have to be paid for by the symposium budget. Fortunately, it was an error. I don't know what I would have done had they insisted; there was no specific mention of this item in the contract and, of course, there were virtually no funds left in the symposium budget.

13.4 Initial budget estimate

If you prepare your budget according to the above suggestions, the skeleton of your expense column may look as follows:

EXPENSES

    I. Projected expenses.. $85 000
   II. Contingency funds (about an optimistic 6%).............................. 5 000
 III. Cost overrun (10% of I and II)... 9 000

Grand total <u>$99 000</u>

Assuming you anticipate 500 fully paying participants, make your grand total $100 000 and divide it by 500. Now it looks as if a basic registration fee of $200 per participant would suffice. But does it really? Probably not, because of the following four categories: (1) costs for accompanists; (2) reduced registration fees for graduate students; (3) special guests (and perhaps their spouses, too) that attend functions free of charge (e.g., university or government officials); (4) at international meetings, the almost inevitable requests from colleagues from 'developing' countries for a reduction or waiver of fees.

For accompanists, the fees can be set so that this budget category will break even. That is if: (*a*) your participants don't sneak in so many unregistered guests, kids, etc. at receptions that you must order additional food and beverages at extra cost; (*b*) unregistered persons do not occupy bus seats which you must pay for individually (see Chapter 11).

Assuming you have only honest participants and accompanists, multiply their total number by the amount of funds per person needed for events in which they participate, add the expenses for whatever else you plan to provide for them (e.g., buses and special guides for shopping trips), and divide the total amount by their total number. Add 10% or more if you are not charging fees for children below a certain age.

With graduate students, you may not create a deficit if you allow them to attend the meeting without buying the proceedings. In other words, if they pay the registration fees for regular participants minus the price for the book. If you charge $200 (regular fees) and subtract $80 for the proceedings, they would have to pay $120. It is more likely, though, you will have a heart and set their fees to a total amount below $100. Since, at social functions, graduate students show a healthy appetite and thirst, a large number of graduate students means a deficit in this category.

Of course, the expenses for special guests, and cancellation of registration fees for some participants from developing countries mean a further deficit. For special guests, this may be a minor amount, mainly due to their presence at receptions and dinners. In the calculation given above, it would probably be taken care of by the $1000 which you added to make it a round grand total of $100 000.

With the fourth category, you may run into major problems, both moral and financial. If you do not have funds earmarked for participants from developing countries, any financial commitment to them has to be paid for from your 'regular' budget. In our example, that means $200 for each person (unless you exclude them from social events and do not buy proceedings for them – which would be terrible). For more on this problem, see Section 15.2.

Thus, at some meetings, the registration fees for regular participants may be strongly affected by expenses in categories (2)–(4); unless, of course, you are assured of special funds for these categories. Normally, this will not happen before you need to make the final decision on registration fees.

There are certainly other ways to project a budget. However, keep in mind that the anticipation of a 10% cost overrun provides you with a reasonable margin of financial security.

13.5 Excess funds

Sometimes, organizers are blessed with a budget surplus. As the following suggestions show, there are simple ways to dispose of the wealth:

(1) Transfer funds to your society's account. Earmark them for the budget of the next meeting, or any other worthwhile cause.
(2) Increase the travel support of participants, especially of younger ones or those from foreign countries.
(3) Buy copies of the proceedings for participants who were not eligible for the book because they paid reduced registration fees (e.g., students).
(4) Buy copies of the proceedings for colleagues who could not attend for reasons of health or money.
(5) Give a bonus to your volunteers and/or assistants who probably deserve it, anyway.
(6) Return restricted funds that cannot be spent in one of the above five ways to the sponsor (which, in this case, is probably a government agency).

14

Fund raising: some would rather see their dentist

14.1 Basic strategies

I have never met a colleague who loved fund raising. However, since the organizer of a scientific meeting is unlikely to escape it, you should consider the following points:

(1)   Whom to ask.
(2)   How much to ask for.
(3)   What you can offer in return.

It is amazing how many people approach potential sponsors like children writing to Santa Claus. Even worse, they send more or less the same letter to anyone they consider a potential sponsor. This shotgun approach is senseless. Remember that you are dealing with people or committees with differing interests. Make sure that you tell them what they want and need to know. This requires carefully written, custom-tailored letters and applications, and a concise style.

The first job of a good fund raiser is to identify sources worth contacting. It may pay to ask colleagues who have recently run a meeting similar to the one that is envisioned. Usually, these colleagues have no interest in fund raising for the time being; thus, they may be willing to share their experiences. Perhaps, they will not disclose all the details, but they may provide names and addresses of potential sponsors. Another way of finding sources of support is to look up the acknowledgments in the programs, abstract volumes and/or proceedings of recent meetings. Last but not least, the advertisements of firms in scientific journals often give a clue.

Once you have identified potential sources of funds, it is crucial to find out who to talk or write to. Some corporations have such a turnover of their management that a 'Who is Who' edition from January may be largely obsolete by July. Since you must be sure to approach the right person, have someone call his or her secretary to confirm his or her current status, correct name (remember how much you hate to see your name misspelled) and address. Then, see if you have a common friend who can soften him or her up. Fund raising by phone or letter has become rather pointless, unless the potential sponsor is either personally interested in supporting you, or can be convinced that your meeting will benefit his business.

In the world of business, appearances are important and money goes where money is. Don't make a phone call without involving your secretary or someone playing that role. When you write a letter, display (but don't pull!) rank. A tactful letter from the president of a symposium is more impressive than one signed by an obscure 'Assistant Professor of Genetics.'

Do not expect a corporation to give funds unless you can offer something tangible in return. Some firms have in-house review committees to handle grant applications. Do your homework and find out whatever you can about them. Relate your application to their specific interests. Your chances of being funded depend on both the interest and influence of your supporters on these committees.

In contrast to the situation in the business world, it is not advisable to play status games with government or international agencies. Usually, they have limited funds and no reason to grease a fat goose, so it would be unwise to try to use status to impress them. In general, the staff of government agencies are helpful and will advise you. Probably, they will also mail you application forms and other pertinent material.

When applying for funds, be it from a private or government sponsor, unrealistic requests are doomed. This means that you need to know how much you should request. In the case of private businesses, this can be difficult since the amount may vary with their interest in a meeting. A local firm may give you a one-time token amount because you are a neighbor, while most of their grant funds go, year after year, to far-away organizations that are important for their business.

How can you learn about the funding policies of a private enterprise? Well, find out if you, or a friend, know an insider who will advise you. Furthermore, as suggested above, try to identify colleagues who have recently dealt with potential sponsors.

What can you offer private corporations in return for their support of your meeting? Obviously, a lot if your participants endorse medications or products that are manufactured by your sponsors. Your bargaining position is weaker when your meeting is dealing with work that may lead to new medications or products of commercial interest. You are often reduced to asking for a handout when you request funds for a meeting on strictly basic research.

However, even for a conference on basic investigations, there is some hope for support. Don't rely just on industry. Instead, remember that your meeting (*a*) brings money to the local economy, (*b*) may provide a good chance to advertise or sell books, instruments and chemicals, and (*c*) may enhance the prestige of the host institution. Of course, the bigger your meeting, the stronger your clout. Thus, try to get: (1) free receptions from the local government, local enterprises (e.g., breweries), your congress hotel and exhibitors; (2) a sponsor for the usual items handed out at the registration desk (e.g., a bank or travel agency); (3) free services and meeting rooms from your college or university.

14.2 Preparation of formal applications

When you submit a formal application for funding, be realistic. The program officer of a government agency and/or the members of a review panel could be smarter than you; and collectively, they may have more experience. Thus, don't try to hide 'fat'; they are likely to find it, and they hate to be taken for fools. If, for instance, you apply for an expensive nominal air fare instead of the discount rate everybody uses, you may not make extra money; rather, you may get nothing. Somehow, people are extremely sensitive when it comes to travel expenses.

Perhaps, the worst approach to travel support is applications that request an unreasonable amount for clerical work or overheads. Let's take a not unusual example. You ask for $10 000 in *direct* costs to support members of your society who wish to attend a meeting abroad. With $10 000, you can support 20 members with $500 each. However, you may have to set up a committee to choose these 20 people from a total pool of 30 applicants. Of course, you could support all 30 applicants with $15 000, the maximum amount mentioned by your potential sponsor. Unfortunately, though, your institution charges 50% in overheads. Neither a private firm nor the reviewers of a public foundation (perhaps with the exception of the US National Institutes of Health) will look favorably upon $5000 in overheads when it means that 10 out of 30 participants cannot be supported. Your chances for funding will be further reduced if you must include $1000 in direct costs for clerical work. The amount left for the actual purpose of the application will then shrink even further, to $9000; that comes to 60% of the total budget, and travel support for only 18 out of 30 interested colleagues. Would you, as an outside reviewer, recommend funding?

Fortunately, there are ways to avoid this kind of situation. Either you apply via your society, provided it charges reasonable handling fees; or you, the organizer or president of the meeting, apply directly with no intermediary at all. The latter approach is usually feasible when dealing with private sponsors and international organizations.

How would a reasonable budget look in our case? You ask for a total of $15 000, including $600 for clerical expenses, to support 30 participants. Instead of $500, you allocate only $480 per person, and your budget is as follows: 30 × $480 = $14 400, plus $600 for clerical expenses.

The advantages of this approach are obvious: (1) no waste of time and money for committee work; (2) all 30 colleagues receive support; (3) no ill feelings of colleagues who were not funded; (4) and, most importantly, your grant application is realistic.

Depending on the distances traveled by the participants, you can modify your budget to compensate for the varying expenses. This may be fairer than a standard amount, and it may be more favorably received by potential sponsors. If, for instance, the meeting is held in California, the support for participants from the North American continent can be allocated so that people from New England and adjacent Canadian provinces receive more money than their colleagues from the Midwest. In Europe, I applied the following scheme for Europeans: (1) foreigners, the highest amount; (2) natives of the country where the meeting is held, intermediate amount; (3) participants from the immediate region of the meeting site, the smallest amount.

What may the reviewers of your application look for? For one, *originality* and *timeliness* of the meeting, and presentation of *novel ideas*. If your preliminary program supports these claims, your application will stand out from the deluge of others that promise more of the usual. Second, new *breakthroughs*. It is easy to convince reasonable people that scientific or technological breakthroughs may save large amounts of time and money, and that they may lead to new frontiers. Third, depending on the prospective source of funding, it may be important to point out the potential *human relevance* and/or *economic benefits* of the issues considered at the meeting. But don't embellish the truth too much; that may cause an adverse reaction. Fourth, provide a meaningful participation of *younger scientists*. Don't just lower the registration fees for them; provide evidence that they can discuss their research with people who will be helpful to them. One of the best ways to involve younger scientists in the program of a meeting are Socratic Workshops (see Section 3.1.4.3). Fifth, if you are applying to a US government agency, remember that they may need an assurance that *women* and members of *minorities* will receive special consideration. Sixth, if it is an international meeting, point out the importance and *benefits for the country* whose nationals are to be supported. Don't be bashful when stating the negative side. Put in plain language that the scientists you want to support will be unable to keep up with the progress in the field, unless they attend the meeting; and how this may affect their future work, and the international reputation of their country.

Finally, abide by these rules of common sense:

(1)   Stick with the format. Make sure that the cover page contains all the requested information, plus the necessary signatures. Stay within the space and page limits. Reviewers may become angry if you exceed page limit, and they may become very, very angry if you cheat: don't use a small font to squeeze in more words per space, and don't expect an excessive appendix to help you. Remember that the reviewers' lifetime is precious to them, and reading your application may not be their idea of fun.

(2)   Don't lie. If you are requested to reveal all sponsors you have already contacted, or are considering asking, write their names down. Especially in the USA, government agencies often consult with each other, and they have

mutual access to certain computer files. If they catch you, your first attempt at 'double dipping' may be the end of your funding history. Also, one of your colleagues may review grants for more than one agency and thus discover discordant information.

If you have the privilege of suggesting reviewers for your application, don't submit names of persons that are clearly inappropriate. Don't name a good friend who is not qualified to evaluate your application. His highly favorable review without substance may be discarded. The same may happen when the text of a review really does not match the high overall rating on the bottom of the form – not at all a rare case. It can be even more damaging when a former mentor reviews the application of a disciple even though this is against the rules of the granting agency. When discovered, it will boost neither the credibility of the applicant, nor that of the reviewer.

(3) Don't be lazy. Frequently, the cover page and/or the list of contents of grant proposals have the wrong format. In the USA, duplicates of applications to the National Institutes of Health are often sent to other agencies without rewriting according to their specifications. If an applicant does not prepare a careful revision, why should a reviewer take the 'dupe' seriously? After all, careful evaluation of the grant may take many hours, perhaps a full day.

Another annoying situation arises with poorly proof-read applications. Unless you have a truly conscientious person to do the job, do it yourself. Check carefully each and every page and every copy you mail out. Yes, even if 20 copies, each with 30 pages, have to be checked. If a reviewer must call the agency because an important page is missing, or only every second page is copied, it costs him time and money, and does not foster a positive attitude.

In a final analysis, what can you offer prospective sponsors of your meeting if commercial advantages are not the issue? Mainly the following: a well-conceived, worthwhile program; a well-written application; and a realistic budget.

15

Allocation of travel support: not much fun either

15.1 Some food for thought

I have been touched by the decency of colleagues who were not well off financially and yet asked to give their share of travel support to younger researchers. On the other hand, I have been appalled by the avarice and egotism of some very illustrious and well-to-do scientists.

The first time I had to allocate funds for a meeting, I called up the invited speakers as soon as I had received the award. Joyfully, I asked the first fellow if several hundred dollars would be helpful for his travel arrangements. The answer was prompt: 'No, not really.' Dazzled, I asked if that meant he would not come. In a diplomatic reflex, he then assured me that he was very happy to accept the money.

Some scientists seem to believe that the rules of a bazaar also apply to requests for travel support. They ask for outrageous amounts hoping that this will garner them the lower amount they are actually shooting for. Some of my friends have become so allergic to this attitude that they look with apprehension at any scientist from certain nations.

Similarly unpleasant is the expectation of some retired scientists to receive lavish travel support for meetings they have attended for decades. Of course, you want them to come, but how can you justify paying for suites for them if you don't have enough funds to pay for beds for outstanding younger colleagues? What granting agency would approve an application requesting preferential funding for retired honorees? If you can afford it, you will probably help the older colleagues by dropping all fees, allowing them a free copy of the proceedings, and perhaps even some additional support. However, if they are used to more, the chances are that they will feel insulted by you; and the wrath of old people is not to be underestimated.

Another problem arises with matching funds. It sounds great when a colleague informs you long before the meeting that he will receive travel support, provided you can match the amount. This may put you in a terrible position. How can you commit money before you know if travel funds will be available at all, or what instructions will be connected with a travel grant? Sometimes, you wonder if there are funds to match at all, or just a manipulator in action.

15.2 Support for participants from 'developing' countries

This issue has become a major headache for organizers of international meetings. For decades, international organizations could be counted on to support scientists from 'developing' nations, and usually some non-restricted funds from other sources (e.g., industry, local governments) were also available. Moreover, organizers could be generous and waive registration fees, or even pay hotel bills.

This situation has changed drastically. International organizations seem less able and/or willing to provide general travel support; some restrict their funding to specific projects, and others provide loans to get the preparation of meetings under way. The organizer may be seriously challenged if he tries to use unrestricted funds or registration fees for participants from 'developing' nations, and rightly so. It would be unfair, if not unethical, to support the latter with funds from the general budget of your meeting. After all, the chances are that most of your money has been contributed by participants from 'industrialized' countries whose compatriots may have paid a considerable portion, if not all, of their travel expenses and registration fees from personal sources.

Occasionally, organizations or committees start a fund drive to raise travel support for colleagues from countries whose currencies are virtually useless. While the intent is noble, this is a touchy issue. People may feel pressurized when their registration form contains a line like 'Voluntary contributions for . . .' This type of fund drive must be decided on a case by case basis, keeping in mind that the situation could become very embarrassing if the drive fails.

Should you really give support to a colleague from a country where modern research is virtually impossible, and refuse it to a graduate student from an outstanding laboratory in Europe, for whom your conference is a unique chance to meet leading Japanese experts in his field? What sense does it make to support participants from 'developing' countries who come to learn new techniques from workers at the cutting edge of science, when the latter do not attend your meeting for lack of funds?

Some of our colleagues don't see it this way. However, the argument that scientists from 'industrialized' nations have an obligation to support those from 'developing' countries has worn thin. Instead, there is a realization that some of the richest nations have mismanaged their economy and natural resources beyond description. When corruption, incompetence and/or cultural attitudes keep such a nation in a permanent state of underdevelopment, how can it expect other countries to routinely support its scientists – especially when it spends lavishly on foreign travel of athletic teams? For sure, politicians set priorities in these countries. However, unless the scientists of certain nations succeed in changing the attitude of their politicians, the quickly widening technological gap will make it pointless for them to attempt modern research at home.

How can we help our colleagues from 'developing' countries to attend foreign meetings? Certainly, one should never miss a chance to raise funds from international organizations. Here, occasionally a little trick may be helpful: if an organization provides only funds for small or special meetings, and yours is a big one, create a specially-named conference that is held during the regular meeting time; or, if necessary, on the first or last day of the meeting. Then, submit an application that requests support for the 'special' conference, and include in the list of participants several colleagues from 'developing' countries.

On the other hand, there should be no problem justifying travel support for outstanding scientists from any country, provided it is also available to their peers. Usually, this type of support will be limited to certain speakers; and in the case of 'equal quality,' you have the option of giving preference to a speaker from a 'developing' country. Furthermore, in order to help colleagues to secure support from their own country, make sure that they will know as early as possible that they have been scheduled to be speakers, session cochairmen, workshop participants, etc. In this way, you can beef up your official letter of invitation, which they most likely will need for their authorities.

Finally, there may be special sources of travel funds available for members of certain nations, nationalities or other groups. These funds may be provided by compatriots or organizations sympathetic to certain causes. It may not hurt to explore this avenue. However, caution is advised when political strings are attached to the support.

15.3 Setting priorities for support

When allocating travel support, usually, the highest priority will be given to the presenters of Named Lectures (see Section 3.1.1.1); and often, this means that you must commit money before you actually have it. However, when a good deal of your program depends on certain speakers, you have no choice but to assure them support, if they insist on it. If you give support to one, you will have to offer it to all whose lectures are of the same 'weight.' Otherwise, some may be insulted and drop out from your program at a late date, and this could create a chain reaction of problems.

Your situation may be particularly unpleasant if your promised funds do not arrive until after the meeting. This happened to me once with funds from an international agency. In desperation, I set up an *ad hoc* committee to decide on the distribution of the expected funds and handed out vouchers to all who were approved by the committee. The vouchers had hardly been distributed when I became the target of the wrath of all who felt slighted, particularly of those who were not even eligible for consideration. Weeks after the meeting, when the money was finally available, I sent checks together with a preprinted form acknowledging receipt of the money (to be signed by the recipients). Several people did not even bother to return the receipt despite reminders.

Never give support only to some and make them promise to keep it a secret. This would be neither ethical nor smart. People promise a lot, but are forgetful; and many like to boast. Also, remember: once you submit a financial report to a granting agency, you have no control over who is going to read it.

If you are organizing a meeting for a society that is traditionally of great interest to certain industries, you may be showered with financial support. In this case, any advice on allocating funds would be gratuitous. However, the international meeting of a society without major links to industry may provisionally allocate its budget as in the following example:

Plenary Lectures
(1) Two Named Lectures: support pledged by reliable long-term sponsor.
(2) Two Named Lectures: support available from surplus funds from previous meeting.
(3) Six other Plenary Lectures without endowment: funds virtually assured from anticipated registration fees.

Action for categories (1)–(3) Invitation to the ten speakers, committing travel support; however, ask if they can pay part, or all of their expenses, from their research funds. If possible, this matter should be discussed by telephone before the written invitation goes out.

State-of-the-Art Lecturers and participants in key roles
(4) Sixty State-of-the-Art Lectures: travel support will depend on success of grant applications submitted by the organizer.
(5) Sixty participants in critical key roles (workshop leaders, colloquium moderators and colloquium participants): travel support will depend on the success of grant applications submitted by the organizer.

Action for categories (4) and (5) Invite all persons envisioned with the clear understanding that no support can be guaranteed; request that they commit themselves to attend even if no support is forthcoming; ask who is willing to relinquish a claim for travel support, if it should become available.

Participants in other scientific events
(6) Other participants: unrestricted travel support will be unlikely. Possibly, a few restricted, specifically earmarked funds may yet become available.

Action for category (6)   Inform all those who request information on the meeting that travel support for them is unlikely; however, if unrestricted funds should yet become available, then these will be given to workshop participants and presenters of 'Short Communications' and/or Posters.

If you have been informed (or you feel) that restricted funds may be forthcoming, indicate the possibility to those who might be eligible.

In any case, do some soul searching and develop a fair key for the distribution of the funds.

16

Schedule of preparations: from dream to reality

16.1 Estimated time frames for preparations

An evergreen problem with scientific meetings is a belated start of the preparations. The following suggested minimum times for *intensive* preparations begin with the 'date of no return,' i.e., the mailing of 'official' invitations and the first announcement to journals. It is assumed that you have a clear idea of the scope, dates and location of the meeting, have consulted with colleagues on format and presenters, and have ascertained the feasibility (see Chapter 2):

(1) Informal research conferences: 3 months.
(2) Regional (specialized) research conferences:
     (*a*) No application for government or international agency support: 6–12 months.
     (*b*) Government or international funds to be applied for: 1–2 years.
(3) Regional general meetings: 1–2 years.
(4) National or international research conferences
     (*a*) When held at regular time intervals: 2 years.
     (*b*) One-time or irregular events: 3 years.
(5) National or international major meetings: 3–4 years.

16.2 Factors affecting the time frames

If you are counting on financial support from government or international agencies, you must adjust your time frame to suit their *deadlines for grant applications*. Since agencies may have a few, or even only one deadline per year, it will be essential to find out exactly when it is, and what to submit. Usually, these applications are expected to contain a reasonably complete program, which means that you must know what you are going to incorporate in the preliminary program (see following sections).

Other important factors in the preparation of meetings are the deadlines for payments and submissions. Good *deadlines for payments* in the northern hemisphere are January 15–April 1. Christmas spending is over, and summer vacation payments are not yet due; in the USA, though, the infamous April 15 tax deadline may cast its shadow. Other good deadlines are between October 1 and November 15. Payments for the summer vacations have been made, and the Christmas spending frenzy is not yet fully underway; and in the USA, the initial expenses for the children's new school year have also been paid. Apart from the Christmas season, bad deadlines are between May 1 and September 15. During this time are the summer vacations of the northern hemisphere; and in the USA, these are followed in early September by expenses for the new school year.

As pointed out in Section 8.3.2, the worst *deadline for manuscripts* is the summer vacations, unless that is when you are holding your meeting, *and* you know how to make your contributors hand in their papers.

16.3 A stepwise approach for minor meetings

16.3.1 Informal research conferences

The preparation of these conferences may not require more than a few telephone calls, plus arrangements for the meeting room, non-alcoholic beverages and cookies, and perhaps also for lunch, dinner and overnight accommodation. Paperwork can be held to a minimum. However, you may consider confirming your agreements, when made by telephone, in a short letter, and follow up with a phone call about three weeks before the event. If the participants are paying for their meals and board, the expenses of the organizer may be minimal, i.e., mainly for coffee, cookies and soft drinks. These conferences are usually local events with less than twenty participants. They tend to be very fruitful. You may combine an informal research conference with the seminar of a visitor, especially when this allows the use institutional funds for his remuneration.

16.3.2 Regional research conferences and similar meetings

There is a sliding transition from the type of meeting above to formal regional research conferences, which are used in the following as an example for minor meetings. Formal research conferences require a structured program and, depending on the presence and size of an audience, considerable preparation. If there are only lectures and no interactive events, the value of such conferences is limited, but your workload is reduced (see Sections 5.1 and 5.2).

With guaranteed and sufficient support from industrial or other sources, a preparation time of six months may suffice for a regional research conference with an audience. However, if you have to apply for support from government or international sources, twelve months or more may be needed.

The following steps assume that: (1) you include an audience (about 100 persons) in the program; (2) there will be Posters; (3) there will be an abstract booklet, but no proceedings; (4) the meeting lasts at least one full day; (5) you will apply for funds from government and private sources.

1½–2 years before the meeting
(1)   Confirm with the meeting site that your dates and basic requirements can be met. Check for possible problems due to competing or interfering events.
(2)   Develop a program in consultation with colleagues, but do not install a formal committee if you can avoid it.
(3)   Decide on the format of the abstract booklet.
(4)   Invite and confirm your speakers and key participants; in particular, identify good moderator(s) if you plan a Forum and/or Colloquia.
(5)   Find out deadlines and details required for submissions to granting agencies and contacts with potential private sponsors.
(6)   If you think it is worth it, develop a logo, letter head and special envelopes; have letter head and envelopes printed.
(7)   Finalize the program as much as possible. It will be essential for your applications for support.

About 1 year before the meeting
(8)   Check out prospective hotels or motels for your participants (see Chapter 6). Get an agreement on their prices (including taxes!), and arrange that *they* will handle the booking. Information on their prices may be needed for your grant application(s). Unless the conference is held in a hotel where the number of your participants is part of a bargain with the management, there is no need for you to keep track of the bookings (it will be different when you organize a major meeting; see below).
(9)   Apply for government funds (remember the deadline(s)!), and follow with a fund drive from other sources.
(10) Subsequently, mail a brief announcement to pertinent journals and those in charge of society newsletters, and invite inquiries. Remember that it may take several months before your announcement reaches its audience!
(11) Make a list of all potentially interested people in your region (you may find their names in the membership directories of societies, and/or the list of participants of a preceding meeting).
(12) Calculate the level of the registration fee needed to cover your expenses, assuming (*a*) you will not receive financial support (do not overestimate the number of fully paying participants!), (*b*) your socializer will be of the 'no host' (cash bar) type.
(13) Have the detailed program (including: deadlines; rules governing attendance

and financial support; registration fees and mode of payment) typed or printed, and copied for distribution.

(14) Develop a simple questionnaire (see Section 18.1) for those who may wish to attend the conference. Request name and address; phone, fax and e-mail numbers; information of interest in presenting poster; information on overnight accommodation.

(15) Develop a form and information for both abstracts and Posters. See Section 8.2, and Appendixes A and J).

6 months before the meeting

(16) Mail the detailed program to all potentially interested people in your region. Include the questionnaire and a form (and information) for abstracts (if applicable). Emphasize that the deadline for abstracts will be 1 month before the meeting (give precise date); request that the questionnaire be returned by the same date. Also include information on overnight accommodation, and point out that booking will be directly with the hotel/motel.

(17) Set up a bank account and arrange other budget-related matters (see Chapter 13).

(18) Based on the response to your mailing, make the final arrangements for meeting room(s) and equipment, poster boards, assistance, catering, social event(s), overnight accommodation.

(19) Start a continuously updated list of participants.

6 weeks before the meeting

(20) Confirm by phone that your speakers and key people will indeed show up; arrange replacements immediately, if necessary.

2 weeks before the meeting

(21) Prepare the forms needed at the registration desk (including receipts for conference fees, letters confirming attendance), name tags, tickets and information (e.g., list of recommended restaurants with a map). Prepare direction signs and numbers for poster boards (see Chapter 18).

(22) Deadline for abstracts; have abstract booklet prepared immediately.

(23) Develop detailed schedules for the staff (see Appendix L), and lists of necessary equipment for registration desk and meeting rooms (see Appendixes M and N).

1 week before the meeting

(24) Confirm that all members of your staff will be available as agreed.

(25) Depending on the size of your meeting, alert the recommended restaurants so that they are prepared for your participants, especially at lunch time.

(26) Confirm arrangements for social events.

(27) Ascertain that there will be no unpleasant surprises with the recommended hotels and motels.

3 days before the meeting
(28) Make sure that all items needed for both the registration desk and sessions are available.
(29) Confirm the arrangements for coffee breaks (emphasize punctuality!) and soft drinks for the meeting site.
(30) Print out list of participants and have it copied.

24 hours before the meeting
(31) Confirm once more that all necessary equipment and supplies, including registration desk and chairs, and pertinent items (electrical outlet!), bulletin/ message board and poster boards are available and functioning. Do you have enough slide trays or carousels and a 'slide viewer'? Is there a trash receptacle for the registration desk, and if self-adhesive nametags are used, another one?
(32) Rehearse the individual roles with your staff.
(33) Prepare name tags for all anticipated participants.
(34) Prepare registration kits for participants that are coded: (*a*) 'has paid' (if preregistration fees were requested); (*b*) 'payment due upon registration'; (*c*) 'message inside.'

2 hours before the meeting
(35) Make the final check of all arrangements at the meeting site. Does the registration desk really have all the necessary items and supplies; in particular: (1) Are all registration kits available and properly coded? (2) Is there enough change for larger bills?
(36) Put up direction signs so that your participants will easily find the meeting rooms (the more signs, the better!). Make sure the bulletin board is in the right place.

1 hour before the meeting
(37) Make sure your staff at the registration desk, the guard at the door, trouble shooter and projectionists (ready to accept slides and put them in labelled trays or carousels) are at their places.

After the last scientific event of the day
(38) Take all registration material, computer and accessories, money and checks, forgotten slides and other items to a safe place.

If your meeting lasts two or more days
(39) Go through steps (37) and (38) again.

(40) Take unneeded cash and all checks to the bank (before it closes!).

Day after the meeting
(41) Make sure all borrowed or rented items are returned.
(42) Pay your staff, if no volunteers could be found.

2 or more days after the meeting
(43) Make sure that there is enough money in the account, and that checks have been credited to your account; then start paying bills.

Weeks to months after the meeting
(44) Close your meeting account when you are sure that you have met all financial obligations.

Comments It would be naive to expect that the above 44 steps apply to smaller meetings exactly as given. Each meeting has its own momentum, and surprises are the rule, not the exception. The steps given here may provide an outline for your preparations, and quite a few reminders.

A closer look at the preceding steps also shows that a formal registration with fees creates a chain reaction of chores. Obviously, you would save considerable time, effort and funds if you had the limited, but sufficient, financial support needed to skip all registration procedures, and to provide refreshments plus a few minor items (such as 'stick-on' name tags, general information, program and perhaps abstracts) at no cost. Which only proves that poverty doesn't come cheap.

16.4 A stepwise approach to national or international major meetings

The following suggestions assume that you are preparing a major international meeting (about 500 people) that is held at intervals in different countries. The meeting includes a variety of scientific and social events, an all-day excursion, and an accompanists' program. Proceedings will be published. You have ascertained the feasibility of holding the meeting at a certain location and at a certain time. There will be no overlap with competing meetings, and no interfering local events (see Chapter 7). Furthermore, you have some funds for the expenses which will start right away (see Sections 2.2.1 and 13.3). However, your meeting must be approved by an international society.

4 years before the meeting Before you submit in writing, or present in person, your proposal, be better prepared than your competition. This means that you have all important facts at hand, including the following:

(1)   Exact dates and location.
(2)   Available meeting facilities, including poster boards.
(3)   Available hotel(s) and approximate room rates.
(4)   Weather conditions and outdoor activities at the time of the meeting (jogging, swimming, golf, hiking, etc.).
(5)   Typical daily expenses, such as for meals and local transportation.
(6)   Examples of current (better even, anticipated) roundtrip fares to some major foreign cities.
(7)   Cost of airport shuttle and ground transportation.
(8)   Estimated registration fees, including and excluding proceedings.
(9)   Envisioned format of proceedings.
(10) Possible excursions and local attractions.
(11) Opportunities for accompanists.
(12) Safety. Be honest about possible crimes, in particular car thefts, pickpocketing and robberies.
(13) Your plans and chances for fund raising.
(14) Anything else that may make your meeting more attractive.
(15) Your immediate need for funds to pay for the early preparations (e.g., office supplies, postage, secretarial assistance). Ascertain if these funds must be repaid later.

Some societies indulge in extensive approval procedures (even for minor symposia to be held during their annual meeting!). In this case, you may have to fill out proposal forms, which is not a bad idea. However, the best proposal will not help if it goes to a committee that is not interested or familiar with the research to be considered. Perhaps, even 'politics' may cause refusal of your proposal. This can be rather embarrassing for an organizer who has contacted his prospective speakers beforehand. Hence, it is advisable to consult with officers of the society and people who have dealt with the pertinent committee before submitting the proposal.

The following breakdown of preparations tries to be as specific as possible. Flow charts and more general outlines would be more convenient to look at. Unfortunately, though, they don't help when it comes to critical details – which occur almost everywhere in the preparation of a major meeting. As mentioned with respect to smaller meetings, the following steps are unlikely to apply to your meeting exactly as given. Rather, they aim to be a detailed guide that will help you design your own specific schedule of preparations.

3–4 years before the meeting

(1)   Estimate the approximate number of participants and accompanists. This may be easy for an annual or otherwise regular conference, provided your meeting will have a similar program and prices (hotels, registration fees, travel

expenses) to earlier ones. If your meeting does not follow a set pattern, try to extrapolate from recent meetings dealing with similar topics (contact organizers).

(2)   Confirm with the prospective meeting facilities and hotels that your dates and basic requirements are met. Inspect them carefully (see Section 6.2) before signing a contract (see Section 6.1 and Appendix H). Check for possible problems due to competing or interfering events.

(3)   Develop a program in consultation with colleagues, but do not install a formal program and local committees. If you must deal with established committees, do not yet contact them.

(4)   Once your tentative program has been established, set up a *functional* program committee, and if necessary, also a local committee and a 'kitchen cabinet' (see Chapter 10). If required, contact the members of the established committee(s); ask them for advice on the program, and whether they are *personally* willing to raise funds to finance their suggestions.

(5)   Develop logo, letter head and official envelopes, and have them printed (you will need them very soon for your correspondence).

(6)   Decide on proceedings. Select coeditors, determine the format, estimate the approximate number of pages and/or words, and contact publisher(s). For details, see Chapter 8.

(7)   Develop a program outline that includes the types of events, the names of plenary and state-of-the-art speakers, and other key persons (e.g., organizers of satellite symposia, colloquium moderators, workshop and discussion leaders, forum moderators, session chairs). For Socratic Workshops, see Section 3.1.4.3.

(8)   Contact prospective presenters of Plenary and State-of-the-Art Lectures, and key persons.

(9)   Based on the responses, adjust your program accordingly. Consider carefully your financial obligations and do not commit funds (especially travel funds) which exceed the funds available at this time, plus the money expected from registration fees (see Chapter 13). Remove all travel fund bargain artists and other potentially difficult people who revealed themselves during your contacts (see Section 9.3) from their projected key roles. Use financial uncertainty plus the immediate need for firm commitments as an explanation, but avoid confrontations that may cause bad publicity for your meeting.

(10) Decide on satellite meetings.

(11) Announce the tentative program in pertinent journals and in newsletters of societies (invite inquiries of prospective participants). For details, see Section 17.1.

(12) Compile a list of all potentially interested people, based on membership lists of societies and the lists of participants in related meetings.

(13) Develop a simple questionnaire (see Section 18.1) for those who wish to

attend the conference. Request name and address; phone, fax and e-mail numbers; information on interest in presenting poster(s).

(14) Mail an extended version of the tentative program to persons responding to the announcement, and all other potentially interested persons. Invite them to inquire for forthcoming details in the preliminary program which will appear about 1½ years before the meeting. For details, see Section 17.2.

(15) Set up filing and monitoring systems for correspondence, addresses, receipts of abstracts, fees and other financial transactions. No specific advice is given since the continuously improving computer programs would probably make it obsolete within months.

2 years before the meeting

(16) Contact granting agencies and other potential sponsors (Chapter 14), and make plans so you can meet their deadlines.

(17) If indicated, inquire whether a society or an international organization will lend you more money for your preparations. However, remember that you are *borrowing* money!

(18) If necessary, contact professional exhibition specialists (see Section 3.4 and Appendix H).

(19) Make preliminary arrangements with airlines, tour operators and other persons, places or agencies (see Section 13.2).

(20) Decide on deadlines for abstracts, registration, payments and manuscripts (see Chapter 8 and Section 17.2). If possible, integrate deadlines with those of your hotels (check your contract!).

(21) Design: (*a*) forms for abstracts, registrations and receipts (see Chapter 18); (*b*) special letters of invitation (see Section 3.1.1.4 and Appendices C, D, E, and F).

(22) Develop the preliminary program, confirm hotel rates, and determine registration fees (assuming that you will not receive financial support other than congress fees).

1½ years before the meeting

(23) Mail the preliminary program with all relevant information (see Section 17.2) to everyone requesting further information. Remember that this mailing will be very important. It will strongly influence the decision of many, perhaps most, people who are considering attending your meeting.

(24) Begin fund drive from private sources.

(25) Prepare grant applications and submit them for the appropriate deadlines to government and international agencies.

6 months before the meeting

(26) Mail: (*a*) a circular with updated information, a reminder of deadlines, and

a form confirming attendance to all prospective speakers and 'active' participants (see Appendix O); and (*b*) detailed instructions to colloquium moderators and workshop leaders (see Appendices D and F).

(27) Based on the response to the mailing, make the final arrangements for meeting rooms, equipment, poster boards, assistance, catering, accompanists' program, social events and excursions, overnight accommodation.

(28) Set up a bank account and arrange other budget-related matters, such as official handling of your budget by your university (see Chapter 13).

(29) Select the printshop for programs, abstracts, name tags, tickets, signs, stickers for luggage, etc. (see Chapter 18).

4 months before the meeting

(30) Deadline for receipt of abstracts, early registration payments and special guestroom reservations. Contact abstract and payment delinquents who have been scheduled for key roles. Ascertain (by phone) what is going on, and replace them instantly if necessary.

(31) Invite (by phone) introducers of Main Lectures (if envisioned). Follow up with a confirmatory letter. Select them from the registered participants who are likely to receive travel support (e.g., state-of-the-art lecturers). Otherwise, you may have to bait them with some form of remuneration (e.g., dropping of registration fees plus free attendance at the banquet).

3½ months before the meeting

(32) Decide if you wish to have the program and abstracts in one brochure: weigh convenience of combined information vs the weight (see Section 17.4). Have abstracts and final program printed instantly.

3 months before the meeting

(33) Mail abstracts and final program (plus stickers for luggage) to all those who have met their financial obligations (to overseas participants via air mail!). If your abstract volume is separate and heavy, consider holding it for inclusion in the registration kit that is picked up at the registration desk. In this way, you may save on postage; however, if the abstract volumes then have to be sent to a country with a customs barrier, beware (see Section 6.3)!

(34) Prepare further forms (e.g., for confirmation of participation at the meeting), name tags, tickets, signs and numbers for poster boards; have them printed (see Chapter 18).

(35) Order portfolios (with your meeting logo, if you are paying for them), pens (with name of the meeting), order writing pads (with logo, if not too expensive) and everything else you wish to include in the portfolio (except for the list of participants).

(36) Finalize arrangements for accompanists' program and excursions.

2 months before the meeting

(37) Mail a reminder concerning the deadline for manuscripts for the proceedings (see Section 8.3.2).

(38) Ascertain that your sponsors will indeed provide funds as promised. In particular, make sure that sponsored receptions can be held as planned.

(39) Check poster boards (number, condition, size). If repairs or substitutions are necessary, act immediately.

(40) Prepare the daily schedule for your staff in detail (see Appendix L).

2 weeks before the meeting

(41) Prepare a list of items for both the registration desk and the sessions. Make sure that all of the following will be available: (*a*) items needed for the registration kits; (*b*) items needed for the meeting office and registration desk; (*c*) a bulletin/message board; (*d*) a preview room where speakers can check their slides (using 'slide viewers' or miniprojectors) before handing them to the projectionists (optional; however, at major meetings it will cut down considerably on misoriented slides, wrong slide sequences, etc.).

(42) Confirm that all members of your staff will be available as agreed.

(43) Confirm all arrangements concerning projection and audiovisual equipment; make sure that the rooms will be dark enough for projections.

(44) If meeting rooms are separated by moveable partitions, make sure that the partitions have no defects; i.e., that they close completely (see also Section 6.2.1.2).

(45) Confirm arrangements with restaurants for evening workshops.

(46) Confirm that the luncheon tickets are reasonably priced, as previously agreed. If there are any changes, alert neighborhood restaurants so that they are prepared for your participants.

(47) Confirm the details (from special diets to music and transportation) of the arrangements for social events.

3 days before the meeting

(48) Make sure that all items needed for both registration desk (including chairs, electrical outlet and trash receptacle) and sessions are available.

(49) Confirm the arrangements for coffee breaks (emphasize punctuality) and special set-ups (e.g., soft drinks and/or bar for poster area).

(50) Print out a list of participants and have it copied.

(51) Prepare name tags for all anticipated participants.

24 hours before the meeting

(52) Confirm once more that all necessary equipment and supplies are available. Do you have enough slide trays or carousels and 'slide viewers'?

(53) Rehearse their individual roles with your staff.
(54) Prepare registration kits that are coded: (*a*) 'has paid;' (*b*) 'payment due upon registration;' (*c*) 'message inside.'

2 hours before the registration starts
(55) Make the final check-up of all arrangements at the meeting site. Does the registration desk really have all necessary items and supplies (including change for larger bills)?
(56) Put up direction signs (the more, the better). Make sure the bulletin and message boards are in the right place.

1 hour before the meeting
(57) Make sure your people at the registration desk, the guards at the doors, trouble shooters and projectionists (ready to accept slides and put them in *labeled* trays or carousels) are at their places.
(58) If manuscripts for proceedings are due, place yourself or a good 'enforcer' near the registration desk ('no manuscript, no travel support').

After the last scientific event of the day
(59) Take all registration material, money and checks, forgotten slides and other items to a safe place.
(60) Go over the manuscripts (share the work with your coeditors) before the event(s) of the evening. Check for major problems such as wrong format (fonts, spacing of lines, indentations), excessive length, missing pages, poor illustrations and tables, wrong style of citations.

The following day(s)
(61) Go through steps (57)–(60) again. However, take all unneeded cash and checks to the bank (before it closes!).

Day after the meeting
(62) Make sure all borrowed or rented items are returned.
(63) Pay your staff.

Days to months after the meeting
(64) Make sure that there is enough money in the bank account, and that checks have been credited to the account; then start paying outstanding bills. However, keep the bank account and budget administration active.
(65) If manuscripts must be reviewed, make sure that reviewers do their job. Before mailing the papers, call them by phone and ascertain that they are willing to return the papers by your deadline.
(66) Call up manuscript delinquents and find out if it is worth waiting for their

submissions. If this would be too expensive, mail or fax a strong letter (see Appendix K).

(67) Prepare manuscripts, index and other pertinent items for the proceedings and mail them to the publisher.

(68) Try to complete all financial transactions, except payment for the proceedings that will be mailed by the publisher to the participants (if the proceedings were included in the registration fees).

(69) Prepare reports to granting agencies.

(70) Arrange for speedy proof-reading of the proceedings, if necessary. Think of vacation times!

(71) Mail copies of the proceedings to sponsors and other 'special' recipients.

(72) Close the bank account when you are sure that you have met all financial obligations.

17

Announcements, programs, and related information: clarity pays

Smaller and specialized meetings may only need one or two announcements, the second one containing the final program. For major meetings, especially international conferences, more announcements are advisable. They are outlined in the following. Note that the 'preliminary program,' which is important for the turnout at the meeting, requires some detailed information. If possible, set up World Wide Web pages and keep the information updated. Perhaps, you can add the abstracts later on. The points discussed in this chapter apply to a major international conference. They may serve as a checklist from which to select the relevant ones for your specific meeting.

17.1 First announcement

The first announcement should be mailed as early as possible. It may be mailed in two versions: (1) A brief 'summary' of the 'tentative program' which goes to scientific journals and secretaries of societies for inclusion in society newsletters. (2) A somewhat more specific outline in the form of a flier that will be mailed to prospective participants, institutions, etc. If your meeting is one of the regularly held conferences, the flier may not be necessary if the 'summary' mentions the date at which the 'preliminary program' will be mailed.

Since you want the 'summary' published without charges, it should contain a maximum of information, with a minimum of words, on the following:

(1) title of the meeting;
(2) precise location;
(3) precise dates;
(4) focus of the meeting;
(5) scientific program;
(6) rules for participation;
(7) date of further information;
(8) address for inquiries.

Example:

> 'The Third International Conference on Surplus Livestock' will be held at the
> OCEANVIEW HOTEL in Bongout, Lavachia, on March 19–24, 2001. The focus
> is on the economic and ecological impact of non-competitive livestock production. The
> program includes invited lectures, symposia and workshops. The meeting is open, and
> poster presentations are invited. A preliminary program will be available in May 1999
> from TICSL, Faculty of Science, University of Chuleta, Chuleta 12345, Lavachia.

If the editor of a journal insists on a shorter announcement, you can reduce it to
two sentences:

> 'The Third International Conference on Surplus Livestock' will be held in Bongout,
> Iraq, March 19–24, 2001. A preliminary program will be available in May 1999 from
> TICSL, Faculty of Science, University of Chuleta, Chuleta 12345, Lavachia.

Note that no organizer's name is mentioned. The reason is simple. It is assumed
that: (*a*) there will be great international interest in the meeting; and (*b*) most
prospective participants have an idea from the preceding two conferences how the
meeting will be run. By not mentioning names, you will cut down on early influence
peddling (which may soon make your life miserable), and a lot of letters inquiring
about travel support (which you are not sure about at this point). Of course, if it
makes you feel good, you can add your name to the announcement and pay the
price.

If the conference is the first meeting of its kind, or if you are worried about
sufficient interest, it would be better to beef up the announcement with the names
of the organizers and/or committee members. Since journals may not like this kind
of extensive advertising, it may be best to ask them to publish an announcement
with the organizer's name, and invite inquiries.

Example:

> 'The First International Conference on Surplus Livestock' will be held in Bongout,
> Iraq, March 19–24, 2001. The focus is on the economic and ecological impact of
> non-competitive livestock production. The meeting is open, and poster presentations
> are invited. Inquiries to: Prof. Isidore Harmlos, Faculty of Science, University of Chuleta,
> Chuleta 12345, Lavachia.

This allows you to mail a flier in response to inquiries. Design the flier so that it
has a section (postcard format) which can be returned to you with some information
on prospective participants (see below).

If the reaction to the announcement in journals is below your expectations, much
will depend on an attractive flier that can be mailed to individual members of
societies and others with a potential interest in your meeting. This puts you first in
the role of designer and public relations manager, and then of payer (for the printer's

bill and postage). In addition, you may have to apply diplomatic skills to obtain the lists of members of some societies.

Suggestions for a flier with six sections (the commonly used format), printed on postcard-quality paper are:

Section 1   Title of the conference; logo; location and dates. This section should be attractive; use large letters, interesting typefaces and – budget permitting – some color.

You may not need to develop your own logo if the chamber of commerce or the local office of tourism can offer you something with a local flair. For example, at a meeting in Malaga, Spain, the birthplace of Pablo Picasso, we selected his pigeon with the olive branch.

Section 2   Focus of the meeting; brief remarks on the venue and local attractions; names of the organizers and committee members.

The description of the focus gives you the opportunity to act like a smart journalist: present it as a 'blurb,' i.e., in one or two sentences that attract the reader. You have seen blurbs in many journals, under the headline. So, don't write the dull sentence you would use for the free announcement in a scientific journal, as given above ('The focus is on the economic and ecological impact of . . .'). Rather, use something like this: 'Focus: Novel ways of replacing government-subsidized, outdated breeding of livestock with superior methods that improve the income of farmers and protect the environment.' Now you have a string of 'buzz words' that should get the attention of government agencies, researchers in livestock and related sciences, farmers, the meat and related industries; and last but not least, environmental scientists and amateurs. Of course, you can also use it if you place paid announcements in scientific and other journals.

If a sponsor pays for the flier, you may use the bottom of this page for a brief acknowledgment, for example, 'Lavachia Airline has been appointed the official carrier of the conference.'

Sections 3 and 4   List briefly the following: types of lectures (if available and space permitting, also the name of the lecturers); details on participation in workshops and other events (if available and space permitting, also the names of the workshop leaders and other participants in key roles); invitation to present posters; envisioned proceedings; information on local weather; availability of accompanist programs. Emphasize that *all* inquiries should be sent to you; otherwise, early confusion is guaranteed.

Insert on the bottom of one of these two sections the address for further information as given in Section 6 of the flier, but add your fax and/or e-mail number (think again before giving your telephone number as well). This information may be important for prospective participants because Section 6 will be mailed back to you.

Section 5   On top of this section, start with this or a similar sentence: 'Please send the second announcement with further details to the following address (type or print, and capitalize your family name).' Leave some space and/or draw a distinctive line underneath before requesting information on the prospective participant. Keep in mind that there will be little space per line if this section is designed in the usual way (about 3.5 inches = 90 mm on US standard size paper). Hence, use the whole width for the participant's answers. Your questions ('name' etc.) should be inserted in small letters underneath the pertinent lines, and the distance between lines at least $\frac{3}{8}$ of an inch (= 9 mm): first line – family name and title; second line – given name(s); third line – academic or other rank; fourth and fifth lines – affiliation; sixth line – street address; seventh and eighth lines – city, state or province (if applicable), postal/zip code, country; ninth line – telephone number; tenth line – fax number; eleventh line – e-mail or other connections.

Depending on your plans, you may also ask a question concerning the prospective participant's interest in certain events, perhaps as follows: 'I am interested in the following workshop(s).' The answer requires an additional line or two. When you now check the layout of the section, you will note that you have used up all available space.

Section 6 (reverse of Section 5)   Your own address. When separated from the rest of the flier, Sections 5 and 6 should constitute a postcard. Indicate where to cut it off from the rest of the flier.

17.2 Second announcement

Usually, the second announcement contains the 'preliminary program.' Remember that people may have to choose for financial reasons between your meeting and competing ones. Hence, *precise* and *detailed* information on the program, travel support and general aspects is essential. The preliminary program should include, apart from the name of the meeting, logo, committees and their members, and the address for correspondence, the following:

17.2.1. Summary of deadlines

Example:

(1) Submission of abstracts
    (*A*) Main conference: December 1, 2000
    (*B*) Satellite meetings: December 1, 2000
(2) Submission of manuscripts
    (*A*) Proceedings of main meeting: March 19, 2001
    (*B*) Proceedings of satellite conferences: to be announced
(3) Advanced registration
    (*A*) Main conference: November 15, 2000
    (*B*) Satellite meetings: December 15, 2000
(4) Lodging reservations (for details, see below)
    Hotels (early rate):   November 15, 2000
    International House (early rate):   November 15, 2000
    Excursion to Mount Harmony: November 15, 2000
    Airline discount (35%):   'First come, first served'

This example tries to make several points: (1) The organizer of the main conference has done his job and set reasonable deadlines for abstracts, manuscripts, and advanced registration. (2) He has chosen November 15 as the deadline for early payments plus abstracts of his meeting. Since the conference has been scheduled for late March, this date is probably the best time for both payments (about four months before the conference, and before the Christmas shopping time: see Chapter 7 and Section 8.3.2) and submission of abstracts. Thus, a single deadline cuts down on potential confusion and allows simultaneous mailing. (3) The deadline for manuscripts for the proceedings is the first day of the meeting. This allows the organizer and his fellow editors to go over the bulk of the manuscripts and contact the contributors for corrections immediately (see Sections 8.3.2 and 16.4). (4) The example assumes that the conveners of the satellite conferences were difficult to work with. They set a bad deadline for payments (at the height of Christmas shopping) despite the organizer's warnings, and they were unable to make arrangements for the publication of proceedings of their own conferences at the time of the second announcement. This may be mixed blessings for the organizer: their deadline for payments may create confusion; on the other hand, their incompetence may prevent publication of bad proceedings which might overlap with the publication of the main meeting.

17.2.2 Mode of payment

At major international meetings, certain people will routinely try to get around payments of any kind, including fees for late registration. When they approach you, it may help to point out that: (*a*) the forms spell out the rules clearly; (*b*) their peers abide by these rules; and (*c*) you must be fair to all. An example of a form that clearly states the rules for registration fees is given in Appendix P. The accommodation form in Appendix Q emphasizes the rules for refunds in case of cancellation.

17.2.3 Program information

(1) Focus. Give more details on the areas to be covered than outlined in the tentative program. Sometimes, it may be useful to emphasize what is *not* wanted. For example, a meeting on comparative physiology may not wish to consider work on laboratory animals (e.g., rodents, dogs, cats, chickens); or a meeting on clinical aspects of diabetes mellitus may not be interested in basic molecular studies.

(2) Location. *Example*: 'The meeting will be held at OCEANVIEW HOTEL which is located in downtown Bongout.'

(3) Dates. Don't make a common mistake of organizers and write: 'The meeting will be held from March 19–24, 2001.' This leaves prospective participants wondering whether the sessions start on March 19, or on 20; thus, people who may not wish to attend the socializer on the first evening (e.g., for financial or time reasons) don't know if they should book accommodation for the night of March 18/19. The announcement does not tell them either if it is worth staying for the night of March 24/25. In short, people are not told whether they can save money for two additional nights – which may be important in their decision to attend your conference. Equally confusing are announcements of postconference excursions and satellite conferences that do not give the precise dates and times (see below). Remember that this can be critical, especially when people plan on taking advantage of cheaper airfares.

Example of clear information
Following registration from 2 : 00 pm to 7 : 30 pm, the program begins on Sunday, March 19, 2001, with a reception at 8 : 00 pm. The scientific program ends on Friday, March 24, at 5 : 00 pm. A banquet will follow on the same evening, starting at 8 : 00 pm.

(4) Scientific program.
(*a*) Preliminary timetable. Give a timetable with the names of the events as outlined in Appendix R.
(*b*) Preliminary list of Plenary and State-of-the-Art Lectures. Give a list of speakers with the titles of their talks as suggested by them. By the time of the final program, some speakers may have changed the title more than once (see Section 8.3.6). Mention the time allotted for either type of lecture.

(*c*) List of Colloquia. List the topics, names of moderators and panelists. Describe the possibilities of participation by the general audience (see Section 9.2.2).

(*d*) List of workshops. List the topics, names of leaders and possibilities for participation (see Section 9.2.3).

(*e*) Oral presentations ('contributed papers'). Outline the possibility, if any, of oral presentations. Spell out the rules. Also, if there are, for example, awards for graduate students, mention it here (see Section 3.1.1.4).

(*f*) Poster Sessions. See Section 3.1.2 and Appendix A.

(*g*) Other events such as scientific demonstrations, forums and business meetings. Mention briefly the title and purpose of the Forum, and any unusual matter to be discussed at the Business Meeting.

(5)   Abstracts. See Section 8.2 and Appendix J.

(6)   Proceedings. See Section 8.3.

(7)   Registration fees. Indicate whether the price of the proceedings will be included in the registration fees (see Section 8.3.1). Also, clearly list which social events and other expenses (e.g., bus services) are included for regular participants, accompanists, graduate students, and children. For further details, see Appendix P. You may also consider charging a further increase in registration fees for persons registering 'on site.' This may induce people to submit their fees earlier and allow better planning of the conference. Also, it may reduce 'traffic jams' and handling of money at the registration desk.

Mention that a confirmation of receipt of the fees will be mailed if they arrive before a certain date. Set this deadline so that the confirmations will arrive at the participants' home well before their departure (which may be early during vacation times).

(8)   Accommodation. Since the costs of lodging may be a critical factor in the decision to participate in your meeting, make sure that the information given by the hotel is clear (see Appendix Q for an example); and be very specific in the information in your announcement, as in the following:

Special room rates (including buffet breakfast and taxes) will be available at the 4-star OCEANVIEW HOTEL, and the 3-star LAKESIDE and TWO RIVERS HOTELS. For details, see enclosed brochure. These hotels are located within walking distance of the main shopping areas. All hotels have air conditioned rooms, outdoor swimming pools and tennis courts. The OCEANVIEW is the official congress hotel, and all sessions will be held there. The LAKESIDE and TWO RIVERS are located 200 meters from the OCEANVIEW. The hotels have free shuttle buses leaving the Bongout International Airport every 30 minutes, and the journey will take 45 minutes. Members of the meeting staff will be at the airport during March 18–21 to assist you (look for their signs).

The hotels will assign rooms with a special view on a 'first come, first served' basis. There will be no extra charge for children aged 12 or younger sharing a room with parents (cots will be provided). Early reservations are advised (Use the enclosed *red* Form A.) Reservations cannot be guaranteed after March 1, 2001.

Graduate students will have the opportunity to lodge at the INTERNATIONAL HOUSE of Bongout University. Accommodation is dormitory style (no private bath-rooms). There are two beds to a room, and the overnight price plus buffet breakfast is $20. Note that the accommodation is separate for each sex. INTERNATIONAL HOUSE is located 300 meters from the OCEANVIEW. The airport bus will stop at the INTERNATIONAL HOUSE. A non-refundable deposit of $20 must be received by December 1, 2000. Beds will be allocated on a 'first come first served' basis. For further details see the enclosed *yellow* accommodation form.

Question: Why ask graduate students for a $20 deposit? Answer: If you don't do it, you may be swamped with reservations that are not honored. This could leave you with empty rooms at the time of the meeting.

17.2.4 Satellite meetings

As pointed out above, clear information on the satellite conferences is essential since the participants must make travel arrangements months in advance. Unless you are directly involved in the preparation of the satellite conferences, do not rely on their organizers. Develop an announcement, have it approved by the organizers of the satellite conferences, and include it in your preliminary program. If the organizers can provide you with a flier etc. in time, include it in your mailing, coordinate your announcement with the information on the flier, and pray that you will not be held responsible if something goes wrong later. If you feel uneasy about the satellite conferences, add a tactful disclaimer to the announcement, emphasizing that these conferences are independently run by their organizers. For further thoughts on satellite meetings, see Chapter 19.

Example

There will be two three-day satellite conferences ('Pig Farming' and 'Rhino Protection'). Attendance is free to all participants of the main symposium, and a fee of US $100 per person will be charged to others. There will be no program for accompan-ists; their fee of $10 includes a reception with blue tea and kosher cookies on the first evening.

The Satellite Symposia will begin on Sunday, March 26, at 9 : 00 am at the following hotels: Symposium A ('Pig Farming') at the GOLDBERG PALACE in Potamia; Symposium B ('Rhino Protection') at the CAMELOT in Niniveh. These three-star hotels have air conditioned rooms, and the participants will be charged special discount rates. Accommodation without air conditioning will be available at lower rates (see enclosed information). The 'Flying Carpet Line' will provide transportation to these hotels. Their air conditioned buses will leave the OCEANVIEW on Saturday, March 25, at 10 : 00 am. The distance between Bongout and the two places is 40 kilometers, and the journey should last 10 minutes.

For further information, contact the organizers:

Symposium A: Symposium B:
Dr Ari Crevel   Dr Kunan Peters
Institute of Susiology   Royal Rhino Program
Mount Fertility   Ozone Bay
Ruritania   Ruritania

17.2.5 Postsymposium excursions

Announcements of postsymposium excursions are another opportunity to commit a cardinal blunder. Let's assume the preliminary program contains the following information:

> There will be a three-day excursion to Mount Harmony following the meeting. Minimum participation 6 persons. Payment must be made by November 15, 2000.

This skimpy announcement has been based on one which I recently received from the organizers of an international meeting. It foreshadowed worse to come: three months before the meeting, I still did *not* know whether the trip would take place, or if my payment had been received; nor did I know the precise dates of the excursion. The meeting was held thousands of kilometers from my home, during the tourist season. Thus, I had to book the flight early, without knowing: (1) when to book my return flight (three, four or five days after the meeting?); (2) if I should prepare for the excursion (a field trip); (3) if I would have to book an expensive weekend air fare. Most annoying of all was the possibility that I might learn only upon arrival that the field trip had been canceled, and then have to sit out boring days at the congress hotel instead of visiting an attractive place elsewhere on my return flight.

As expected, the confusing announcement was a forewarning. Suffice it to mention that some participants learned on the evening before the excursion that their reservations (made 18 months in advance) were not to be honored, while others almost missed the flight because we had been told the wrong departure time. Who had created the mess? Employees of a government-run travel agency.

Obviously, the following announcement might be more helpful, and instill more confidence:

> A three-day excursion to Mount Harmony will leave on Saturday, March 25, at 9 : 00 am, and return in the late afternoon of Tuesday, March 28. Minimum participation 6 persons. Payment must be made by December 1, 2000. If the trip has to be canceled, the prospective participants will be informed by December 15, and their fees will be refunded at this time. For details of the trip, see the enclosed flier. We recommend that you spend the night March 28/29 at the OCEANVIEW and book your return flight for March 29. Our special room rates at the OCEANVIEW will remain unchanged until the morning (check-out time) of April 8.

However, note: It is assumed that the enclosed flier mentioned will have detailed information on the trip, including advice on lodging, food and beverages, clothing, shoes and photography.

17.2.6 Accompanists' program

An accompanists' program will strongly affect the number of participants in major, especially international, meetings. If you want to use it to attract participants, trips of general interest are important (see Chapters 4 and 11). Describe in a few sentences the places that will be visited.

Example

> The world-famous bazaars offer unique opportunities to buy both traditional and modern crafts. Watch artisans at work and buy jewelry, leatherware, silks, carpets and many other items at bargain prices.

Also, give information on the daily schedule.

Example

> The buses for the accompanists' program will leave the OCEANVIEW at 8 : 45 am on Monday, March 20 (the bazaars); Tuesday, March 21 (riverboat trip); Thursday, March 23 (archeological museum); and Friday, March 24 (camel farm, with riding lessons). The buses will return after the accompanists' lunch in attractive restaurants (approximately $8 including tips) at 3 : 00 pm. The all-day visit to Alladin's Oasis on Wednesday, March 22 (for details see below) is for both participants and accompanists.

17.2.7 Travel support

This is one of the toughest issues because you may not know at the time of printing of the preliminary program what funds will be available. Depending on the size of your meeting, an indication of optimism in your announcement may result in heaps of requests and inquiries.

To dispell rumors and accusations of unfairness as much as possible, point out in your announcement the order of priority if funds become available (for example, that invited plenary lecturers will be given priority). Indicate in very cautious terms any possibility of further support, and how it will be handled.

17.2.8 Travel discounts

As pointed out in Section 13.2, airlines may give considerable discounts for participants of larger meetings. Inform your prospective participants of your agreement. For instance, whether they will receive a discount via your 'official' travel agency, or whether your 'official airline' will give a discount if they show them a letter – which you will mail to all participants who have registered before a certain date. Give the exact amount of the discount, the deadline for booking, and make it clear that you will mail the magic letter only upon the receipt of proper payment.

17.2.9 General information

(1) The meeting place: Point out the major attractions of the meeting place. Act like a good promoter of tourism, but don't lie. After all, your participants will find out!

Example

> Bongout has many historical sites, and several outstanding museums with ancient and modern art. The National Theater features both folklore and international performances. The new Desert Zoo and the Botanical Gardens have attracted world-wide attention. Exquisite jewelry and crafts are available at bargain prices. The city also has an unusual number of modern restaurants with a large selection of national and international dishes. See also the announcements in the 'Accompanists program' above.

(2) Weather: Give specific details on the probability of rain, daytime and night temperatures, and the possibility of using outdoor swimming pools. Add some advice on clothing.

(3) Airport connections: If it is a big international airport, something like the following statement may suffice:

> The Bongout International Airport is served by many international airlines, including 'Lavachia Airline,' our official airline.

(4) Local transportation:

Example

> A new subway system connects the downtown area with many outer districts. Taxicabs have standard fares and provide reliable service.

(5) Banking: Mention the following: name of national currency and its approximate exchange rate with the US dollar; which foreign currencies are generally accepted; whether exchange of money is possible; availability of banks; banking hours; acceptance of foreign credit cards.

(6) Safety: If necessary, alert your participants of potential problems (for example, pickpockets, car thefts, unsafe districts). See also Section 6.1.3.

(7) Maps: Provide a simple but readable map showing the meeting place, airport, the hotels used by participants, and sites of social activities. If necessary, also provide a second map with the sites of the satellite meetings (Potamia and

Chuleta) and the destination of the all-day excursion (Alladin's Oasis). More detailed maps will be needed for the final program (see below).

(8) Passport/visa requirements:

Example

Citizens of the following countries need only a valid passport: Akkad, Oz, Ruritania, Sumer. Citizens of other countries may need a visa. Inquire at the Lavachian Consulate.

(9) Vaccination:

Example

Though no vaccinations are required, immunization against malaria, typhoid and yellow fever is recommended.

(10) Official letters of invitation:

Example

Personal invitations, signed by the organizer, will be mailed on request. If you wish to have specific phrases or 'buzz words' included, please submit a draft of the expected letter. However, the final form of the letter will be decided on by the organizer. It is understood that the letter is not a commitment of the organizer to provide financial support.

(11) Certificate of attendance:

Example

These certificates will be issued to registered participants on request during the meeting.

(12) Insurance disclaimer: This is a critical matter in places with poor hygiene conditions, and/or where street crime is rampant (see Section 6.1.3). To avoid legal or other unpleasant encounters with participants turned victims, it may help to include a statement, similar to the following:

Every effort will be made to ensure the safety and well being of the participants. However, no responsibility can be taken for any accident, theft, damage or health-related problem during the conference. All participants are advised to carry appropriate personal insurance.

17.3 Final program

Since most participants will carry the final program with them, keep it to a reasonable, convenient size. Don't overload it with irrelevant information; of course, sponsors' advertisements must be included. On the other hand, provide information that (*a*) will be asked for routinely by people who do not wish to carry both the preliminary and final program, and (*b*) is helpful for decisions on non-scientific activities (e.g., restaurants and shopping; see below).

Suggested layout:

(1) Cover page: title of the conference, logo, location, dates, name of honoree (if any), name of organization or society involved, perhaps also a drawing or photograph. If you are austere, you may also list the names of the organizer(s) and committee members on the cover and save a page. Overall, the design of the cover is a matter of taste. Use the inside of the cover for information about the owner of the program. Provide lines for his name, affiliation, hotel room and telephone number.

(2) Page 1: 'Contents of the program.'

(3) Page 2: 'Acknowledgments of sponsors.'

(4) Page 3: Important telephone, fax and e-mail numbers (symposium office; congress hotels; physicians; baby sitting service; police).

(5) Also on page 3: Information about the symposium office (staff and opening times) and persons in charge of equipment and posters.

(6) Page 4: Names of the organizer(s) and committee members.

(7) Page 5: Overall timetable with more details than in the preliminary program (see Appendix B).

(8) Detailed instructions for those in 'key roles' and other participants in the scientific program may be printed on separate sheets and mailed only to those concerned.

(9) Program summary.

Example

Sunday, March 19, 2001
15 : 30–20 : 30   Registration (Oceanview Hotel)
20 : 30–23 : 30   Informal get-together (Oceanview Hotel)
Monday, March 20, 2001
08 : 00–12 : 00   Registration (Oceanview Hotel)
09 : 00–09 : 25   Opening of the conference
09 : 30–10 : 20   John Frog lecture (A. B. Rich, USA)
10 : 25–12 : 15   State-of-the-Art Lectures: Sessions (1A)–(4):

        (1A) Evolution of domestic cattle A
        (2) Effects of 'invertebrate hormones' in chickens
        (3) Bone diseases in domesticated turtles
        (4) Habitat destruction by overgrazing

12 : 15–14 : 00   Lunch break
    ***International Committee on Rhino Preservation:
        Luncheon # 1.
14 : 00–14 : 45   Plenary Lecture (M. Syed Ali, Pakistan)
14 : 45–15 : 30   Poster preview
15 : 30–17 : 00   Poster discussions
17 : 00–18 : 40   Colloquia (A)–(D):

(A)   Development of new cattle races
(B)   Estrogen effects in the 'singing boar' strain
(C)   Improved turtle farming
(D)   New legislature on ranching

18 : 40–20 : 30   Evening break
20 : 30–23 : 30   Welcome party

Comments   (1) In this example events have been scheduled so that there is little overlap between the topics of the State-of-the-Art Lectures or those of the Colloquia. Both are held in four parallel sessions (see also Sections 4.1.1.3 and 4.1.3). (2) The time allowed for the 'opening' of the conference has been held to a minimum. This can be done if the organizer speaks himself; otherwise, he may ride a powder keg. A speaker going ten minutes overtime could make the rest of the morning's program a mess. (3) The topic on 'The evolution of domestic cattle' has been scheduled for more than one session (because of great interest). Session 'B' is scheduled on one of the following days. (4) The lunch break is used for a committee meeting. (5) The Poster Sessions and Colloquia should be related to the topics of the morning.

(10)   The detailed daily program

Example for a morning program:

Monday, March 20, 2001
09 : 00–09 : 30   Opening of the symposium (Moscow Room)
09 : 35–10 : 20   John Frog Lecture (Moscow Room)
              Speaker: Albert B. Rich (University of Fundland, Fundland, Texas, USA): 'The future of livestock in the northern hemisphere'
        Introduction: John Buckless (Poorhouse University, Poorhouse, Pennsylvania, USA)
10 : 25–12 : 15   State-of-the-Art Lectures: Sessions (1A)–(4)
  Session 1 (Moscow Room): Evolution of domestic cattle A
           Chair: A. B. Tasman (Netherlands)
                R. Yeng (Taiwan)

10 : 25–10 : 55   A. Ono (Japan): 'From beer to shabu-shabu'
10 : 55–11 : 10   Coffee break
11 : 10–11 : 40   E. Schmidt (Germany): 'Breeding of superior milk cows'
11 : 45–12 : 15   J. Vacas (Spain): 'The evolution of cryptorchidism in toros'

Comments   Continue as above with the listing of the other sessions featuring
State-of-the-Art Lectures. Note that the 'Opening' takes as much time as a State-of-
the-Art Lecture. On the following days, begin with the Plenary Lecture and follow
with four State-of-the-Art Lectures plus a coffee break (after the second State-of-the-
Art Lecture). In this way, your 'standard frame' for the mornings remains almost
unchanged.

Examples for an afternoon program:

14 : 00–14 : 45   Plenary Lecture (Moscow Room)
                              Speaker: Linus Growling (Margarine Foundation, Palm
                                         Springs, California, USA): 'Goats on suburban
                                         lawns'
                              Introduction: Mauricio Chotacabras (Fundación 'Leche Meren-
                                         gada', Matalobos, Spain)

Poster sessions 1–11

14 : 45–15 : 30   Poster preview without authors
15 : 30–17 : 00   Poster discussion with authors

Poster Session # 1: Special purpose races

(1)   Azar, M. A. (Lebanon): New cattle strains for hot climates
(2)   Britto, S. L., Marques, J. & Blazques, J. G. (Argentina and Brazil): A new
       strain of all-purpose cattle
(3)   Penta, M. & Lombardi, S. (Italy): Indigestion in aging mozzarella cows, etc.

Poster Session # 2: New poultry strains

(1)   Abraham, S. (Hungary): Puszta chickens
(2)   Ivanov, I. P. (Russia): Hen-feathered roosters
(3)   Rodrigues, E. and Portillo, A. E. (Puerto Rico): Egg size in gringo hens, etc.

Comments   Poster Sessions are organized according to topics, and numbered con-
secutively. Within a given Poster Session, list the authors alphabetically. At the
end of the program, all authors and coauthors of all presentations are incorporated
in the alphabetical 'Index of contributors.' For more on Poster Sessions, see Section
3.1.2 and Appendix A.

Colloquia A–D

17 : 00–18 : 40   Colloquium B (London Room): 'Estrogen effects in the "singing boar" strain'
(Organized by P. I. Hamilton, Jamaica)
             Moderator: S. Schmaltz, Israel
            Panelists: P. I. Hamilton (Jamaica)
                     G. Pignataro (Italy)
                     C. D. Pork (Singapore)
                     W. Swinehoe (Ireland)
                     P. Van Baconen (Netherlands)

Comment   For details on Colloquia, see Section 3.1.3.

Evening (Socratic) Workshops 6–12

Workshop 9
   (Restaurant '1002 Nights'): 'The post-ranching ecology of the Great Plains'
             Organizers: U. Graser (England) and V. Grashalm (USA)
            Participants: K. Grasheim (Austria)
                     D. Grasmuck (Switzerland)
                     J. Grasse (France)
                     L. Grassi (Italy)
                     T. Graszki (Poland)

Comment   For details on workshops, see Section 3.1.4.

(11) List with detailed information on social events: Briefly sketch the type of social events; give precise locations and times; advise on dress, if relevant; mention payments, if not previously collected.

(12) Index of contributors: List authors and coauthors, and all other 'active' participants (including colloquium panelists, workshop participants) in alphabetical order, with the pertinent page numbers. If there is a large number of poster abstracts, it may help to indicate them by using italics.

(13) General information:
    (*a*) Location of the congress hotels. Give precise instructions on how to reach the hotels – assume that no assistance from your staff is available. Include advice on how to direct taxi drivers in a few simple words.
    (*b*) Information on parking. Point out exactly how to reach the entrance of the parking facilities (refer to a map given below, if necessary). Mention safety measures. Inform participants of any discount, and how to get it.
    (*c*) Precise location of restaurants used for workshops, etc. Advise on how to get there: for example, four persons to a taxi; or will there be a special bus?

(*d*) Local transportation (taxis, public buses, subway, etc.). Inform on the best ways to get around, approximate prices, and which transportations *not* to use.

(*e*) Car rental. Mention recommended agencies; the availability of cars with automatic shifts (important since some people do not wish to use a stick shift); discounts for participants. Give telephone numbers.

(*f*) Baby sitting. Give details, including telephone number(s).

(*g*) Money matters. List recommended banks and their locations; give banking hours; mention other possibilities for exchanging money (airport, hotels, travel agencies?); where and how to get the best exchange rates (with traveler checks or cash?); acceptance of credit cards.

(*h*) Relaxation. For example, briefly describe the following (and how to get there): beaches; good restaurants, with a note on their specialities; recommended shops; theaters; casinos; amusement parks; golf courses; folklore shows; ikebana and Japanese tea ceremonies; and whatever else is of interest to visitors.

(*i*) Airport tax. This may come as a surprise at departure time after people have spent all their funds in local currency (in some countries, export of local currency is illegal!). This form of robbery can cause problems, especially when long-distance flights across the equator leave late in the evening, i.e., after airport exchange counters have closed.

(*j*) Electric current. Inform your participants what voltage is used, and the type of plug required (US, European or Australian type). Suggest that they buy adaptors in their home country. You don't want to create hysteria when someone with long hair can't use his or her drier on the evening of the first social!

(14) Maps. For smaller meetings, you may need only one or two maps. At a major meeting, provide a plan of the hotel floors with meeting rooms; a map of the downtown area with recommended restaurants and shops, and major attractions; and if indicated, a third map showing important places outside the downtown area. Use the outside of the back cover for the most important map.

(15) Empty pages for notes. Even if you provide writing pads with your conference kits, these pages will be helpful to people who don't like to carry a portfolio.

17.4 Abstract volume

Keep this as simple as possible and of a convenient size. Large, beautifully designed abstract volumes may be impressive, but people hate to carry them around. On the other hand, make sure that the lettering is not too small, especially when it is reduced during the printing process. For further details on abstracts, see Section

8.2 and Appendix J. On the cover page, give essentially the same information as on the cover page of the 'final program,' and use the inside of the cover for information about the owner (as suggested above for the program). However, use different colors for the covers when program and abstract volume have the same page size. For smaller meetings, it may be possible to combine program and abstracts in a single booklet.

Provide a list of contents, e.g., in the following way:

Section I : 12 Main Lectures (pages M1–M6).

Section II : 106 State-of-the-Art Lectures (pages S1–S27).

Section III: 371 Poster Presentations (pages P1–P93).

List of authors (list both senior and other authors in alphabetical order with the numbers of their abstracts).

Arrange the abstracts in alphabetical order by the senior authors' names, and use different colors for the three sections. For the user of an abstract volume, an index of subjects, and sometimes a second one for species, is helpful. Since meaningful indexing is difficult, you may ask the authors to supply one or two key words (provide space on the abstract form; see Appendix J).

17.5 List of exhibitors

If you charge exhibitors (see Section 3.4), it is only fair to list them in the program.

18

Design of forms, nametags, tickets, signs and stickers: how about using common sense?

You have probably come across one or more of the following:

Forms that must have been designed by morons. Lines for some items are too short, while ample space is wasted elsewhere. Some lines are so narrow that typing becomes a hit-and-miss experience. The questions are confusing.

Nametags written in a letter size that makes reading from a normal conversation distance impossible.

Tickets, issued for different events, meals or drinks, that are so similar that you confuse them.

Signs, supposed to direct you to buses waiting at the airport, that can not be found.

Of course, at your meeting, it will be different. You will personally supervise the design of forms, tags, tickets, signs and stickers.

18.1 Forms

18.1.1 General layout

Some common-sense rules for the design of forms:

Rule # 1: Keep the requested information to an 'optimal minimum.' In modern societies, people are overexposed to requests for information. Many have become allergic to this harassment and no longer concentrate when faced with numerous questions, no matter what the purpose of the form. This could become your problem. For instance, when a single missing digit of a credit card number requires expensive overseas calls.

Rule # 2: Ask simple, clear questions.

Rule # 3: Give unambiguous instructions.

As discussed in Section 18.1.2, this is particularly important for international meetings when writing dates, amounts in foreign currencies, telephone numbers and personal names.

Rule # 4: Use a system that double-checks all calculations. Whenever there is the slightest chance of confusion, set up forms so that numbers (days, amounts of money, etc.) are checked horizontally and vertically (see Appendix P).

Rule # 5: Assign the correct amount of space to different items. Typical mistakes are lines which are too short for names, and too few for mailing addresses.

Rule # 6: For international communications, ensure that the written areas can be photographed or faxed by machines used for the shorter US format.

While there is usually no problem with copying papers in the shorter US letter-size format (8½ × 11 inches = 216 × 279 millimeters), it is sometimes difficult to use US format machines to copy the complete information given on the longer foreign stationary. Keep this in mind when designing forms for international meetings, and leave sufficient empty space at top and bottom.

Rule # 7: Use colors if you are working with different forms.

Colors can be very helpful when different forms must be handled, and also when tickets for several events are issued. Perhaps, the best way to design forms for a meeting is to use both different colors and a simple system that uses letters. For example, Form 'A' would be red and use a large capital A in the upper right corner; form 'B' would be yellow and use a 'B' of the same size in the same place.

18.1.2 Potential problems at international meetings

Costly mistakes can happen when it is not clear if information is given in the US way, or in the usual international format. A good example is dates. In the USA, dates are given in the sequence month/day/year. In the rest of the world, it is day/month/year. Consequently, 01/12/98 would be January 12, 1998 in the USA, and December 1, 1998, elsewhere. Thus, whenever providing information on dates, one should spell out the name of the month. If a form uses numbers only, the directions must leave no doubt about the places for days and months.

A second problem arises with the writing of large numbers. $1,000.00 is the way for one thousand dollars in the USA; however, inexperienced foreigners would expect it to read $1.000,00. Also, $100.00 and $100,00 could be confusing. Hence, make sure that any announcement for payments leaves no doubt about the amount that is actually meant.

Problem number three comes with telephone and fax numbers. For a foreigner, especially a non-European, it can be frustrating to have to figure out what to dial when a number is given as follows: 0043 0316 925 1234. He may find out eventually that he must not dial 00, but only the country code, i.e., '43.' Now he will reach Austria, but his call will not go through. Why? Because '0316' is the number of the city of Graz for calls within Austria; for international calls, the '0' must be dropped. With the help of two international operators, he may finally learn the number for calls from outside Austria: '43-316-925 1234.' Keep this problem in mind when designing letterheads for international meetings.

Confusion between first names and family names is common. Always request that, except for signatures, the family name is printed and capitalized, whereas the other names start with a capital letter and continue with small ones. Also of course, a form should clearly indicate which name is to be placed where. For the majority of names, the sequence family name/first name/middle initial is useful. Remember the synonyms: first name = christian name = given name; family name = surname = last name. Obviously, one should avoid the term 'christian name.'

Particular caution is needed with names in certain languages. Lizst Ferencz is the normal Hungarian way of writing the name of the composer Franz LISZT (in his homeland, family names are given first). In Portugal and Brazil, Maria Ferreira Marques would be on a form 'MARQUES, Maria F.' Aldo Carneiro Lopes Neto would be the equivalent of the American Aldo CARNEIRO LOPES 3rd. There is a story of a Brazilian business tycoon who returned deeply insulted to his country because they could not find his hotel reservation. It turned out that he had been listed as 'Mr NETO.' The Spanish name Maria Ferrera Marquez would appear on a form as 'FERRERA MARQUEZ, Maria,' or alternatively only as 'FERRERA, Maria.' Portuguese and Spanish names can create many more confusions; e.g., when a son sometimes, but not always, attaches his mother's name to his family name, or when the full name of a woman is Maria del Pilar SANCHEZ PEREZ DE MONTOYA Y LEON. One time, I offended a mother terribly when I called her little Rosario a beautiful boy; paradoxically, Rosario ('Rosary') is a girl's name. Among the many other problems with names is the proper writing of German noble titles. Martin von Fisch should be listed as 'FISCH, Martin v.'; or alternatively as 'v. FISCH, Martin' and in alphabetical order under the letter 'F'. Karl von der Heid, on the other hand, would be listed under the letter 'V' as 'VON DER HEID, Karl,' because von der Heid is a commoner's name meaning 'from the heath.' The same goes for the Dutch 'van.' Piet van Motten would be listed as 'VAN MOTTEN, Piet.' Chinese names can be a special challenge. 'LIM Chinq-May' may be the common form of spelling in many places. In Singapore, it may be written without the hyphen, which leaves foreigners with the question of which name is the 'family name': 'Lim,' 'Chinq' or 'May'? A mixture of Chinese and western names is not uncommon in Hong Kong and Singapore. How about 'Chen Kee Lin, Peter'? Well, the proper quote in publications would be 'CHEN, P. K. L.' These examples, which could be continued for pages, try to make two points: (1) It is important to list names properly. (2) Don't give up easily when the clerks of a hotel cannot find someone with a 'strange' name; where would you look for 'Donald Schomberg McDonald'?

One more note on names. The unfortunate way many computer keyboards have been designed requires special steps to create the *Umlaute* ä, ö, ü, and several common other features, such as accents or the cedille. When organizing an international meeting, make sure that these letters etc. can be typed. You don't want to annoy Dr Inönü from Ankara and Professor Bärenhüter from Zürich (whose names are mispronounced in anglophone countries, anyway), and you sure want to be nice to Françoise Fenêtre from Paris, and Carmen Muñoz from Madrid.

18.2 Nametags

Even some professional organizers do not comprehend that nametags should be (*a*) informative, and (*b*) readable without contortions. At a typical meeting, you would like to identify three different groups by nametags: (1) those who can help and give information; (2) participants; (3) accompanists. However, don't design saucer-size batches with foot-long ribbons for the members of the organizing committee, or for yourself; that could provoke comparisons with cattle on their way to alpine pastures.

Assuming you can restrain yourself, you may choose only two sets of nametags in different colors. For example, white for participants, and yellow for accompanists. For the staff and the committee, an obvious red rim on a white nametag may suffice. If the members of your committee need more tinsel, attach a red ribbon to their tags. On the top of the tag, have a single line spelling out (abbreviated if necessary) the name of the meeting. Space permitting, include the logo. Since the participants know the name of the meeting, there is no need to use much space for this. However, the most important information, i.e., the name of the person, should stand out clearly so that it can be read from a normal conversation distance. The letters of the name should be at least 0.6 centimeters (about ¼ inch) high, and the full width of the tag should be used, if necessary. Write the name that is used in publications in capital letters; for the other names, capitalize only the first letter. Do not waste space on status symbols; the title 'professor' does not improve anyone's data. The line under the name should give the wearer's affiliation; because of limited space, it may have to be given in slightly smaller letters (abbreviate university 'Univ.' or even 'U', if necessary). Universities sometimes have lengthy names that are little known and thus uninformative. If there is no danger of confusion, avoid 'Johann-Wolfgang-von-Goethe University, Frankfurt a.M.' Simply write: 'University of Frankfurt a.M.,' or shorter even 'U. of Frankfurt a.M.' At international meetings, add at the bottom the person's country of origin in capital letters of the same size as that used for the name.

Recall that there are word-processing programs that can be most helpful in preparing nametags. If your computer at the registration desk has such a program, you will not need an old-fashioned typewriter with large fonts when nametags for late registrants have to be written.

Make sure that your nametags are of sufficient size (at least 5 centimeters high and 7 centimeters long; i.e., about 2 × 2¾ inches), and that they fit their holders. Use only light holders and metal clips (or fine safety pins) so that there is no damage to delicate blouses. Holders should live up to their names and not be 'losers' from which the nametags slip out.

For short meetings, 'self-adhesive tags' with hand-written names may suffice. Even then though, have the name of your meeting printed on the tags if you anticipate that people may try to avoid paying conference fees or other expenses. When you hand out self-adhesive nametags, keep a trash receptacle nearby for the portions that are stripped off. If the participants are to write their names on the tags, provide markers.

18.3 Tickets

Tickets for social events, excursions, etc., should (*a*) be of different colors, (*b*) clearly state that they are for your meeting (use the logo), and (*c*) identify in very large letters the specific use (e.g., 'LUNCH ON MONDAY,' 'BANQUET,' 'TIVOLI EXCURSION,' 'ONE ALCOHOLIC BEVERAGE'). Do not indicate prices since they may change, or differ (e.g., for children).

18.4 Direction signs

There can hardly be too many direction signs. Put yourself in the shoes of an arriving visitor who has never been to your meeting site. Where would the first sign be useful? As he drives into your campus? Where would he need the next one to find the parking lot? How would he find his way from the parking lot to the meeting room? Which signs would help him inside the building? How will he find the cafeteria?

If a lecture room has two doors, and one of them is located next to the speaker's place and screen: can it be locked, and a sign be put up so that people enter through the other door (before trying to break the lock)?

Last but not least, if there are any changes in room assignments from one day to another, make sure that the direction signs are changed accordingly. At one meeting, I missed a lecture because the obsolete signs from the previous day had not been removed.

18.5 Other signs

The most beautiful signs will not help if they are not visible. This happened at one international meeting where the assistants were too shy to raise their signs so they could be seen. They just held them, in the airport crowd, up to their navels. Naturally, the tired participants overlooked the signs and went to their hotels without the promised help. At another meeting, the hired buses at an airport did not have any signs. As a result, the arriving participants took expensive taxis, and the buses returned half empty to the hotels. At one all-day excursion with several hundred people, we returned without two compulsive photographers. Later, they claimed that they could not identify our unmarked buses among scores of others on the parking lot.

18.6 Stickers

At crowded airports and railway stations, stickers on luggage will help you to find participants of your meeting. Stickers may be especially useful when people must be identified quickly because shuttle buses are only permitted to load and unload. Also, stickers may help participants recognize each other; this may allow them to spend waiting periods together, or jointly rent local transportation. To be effective, stickers must be of reasonable size and distinctive. The logo of your meeting on a colorful background may be a good idea. Send the stickers with the last mailing to your participants before the meeting.

19

Satellite meetings: think twice

Satellite symposia can save travel funds for both the main and satellite meetings. On the other hand, they may create problems for the organizer of the main meeting. To avoid this, satellite symposia lasting longer than one day should only be permitted under the following conditions:

(1) They are held *after* the main meeting.
(2) They do not compete for funds with the main meeting.
(3) Their presentations are coordinated with those of the main meeting.
(4) Their budgets are kept separate.
(5) Travel support for participants of both meetings is coordinated.

The first condition is based on the experience that: (*a*) most people feel tired after several days of scientific discussions; and (*b*) a specialized meeting still attracts a good audience when the general interest is waning. If a general meeting follows a specialized symposium, the chances are that tired participants will either leave early, or spend more time in the hallways than in lecture rooms. And you, the organizer, will cringe at these 'lobby lizards' when the audience in some sessions drops to an embarrassing low.

On the other hand, a conference dealing with a topic directly relating to one's research has an enlivening effect, provided there is a chance for discussion, and not just another barrage of talks. Thus, a well-conceived satellite symposium should not suffer from a preceding main event. A good satellite symposium will provide opportunities for personal interaction, such as Workshops, Round Table Discussions with participation of the audience, and Poster Sessions.

The second condition is self-evident, but all too often ignored. An extreme case should make the point. When a satellite symposium was approved in connection with an international symposium, its organizers started a vigorous fund drive from national sources. From these funds, they did not contribute a penny to the main symposium. Thus they became competitors for local and national funds while, at the same time, they were beneficiaries of the international travel funds raised by the organizer of the main symposium.

The third condition is also self-evident. It stipulates that the program committees of the two events closely cooperate so that the events complement each other to mutual benefit.

The fourth condition aims to avoid financial and other entanglements that make book-keeping a nightmare. However, it must be reemphasized that independent budgets should never cause competitive fund raising. Therefore, from the point of view of the organizer of the main event (though not necessarily of the participants), the ideal solution may be to hold satellite symposia of an international conference in a different, but nearby, country.

The fifth condition aims to prevent 'double dipping' by participants.

Overall, the organizer of a scientific meeting should carefully weigh the advantages of permitting satellite symposia. In times of restricted travel funds, they could become a drain rather than a bonus for the main meeting.

20

Checklist of important no-nos

Do not:

accept responsibility without AUTHORITY;
underestimate your EXPENSES;
underestimate the TIME needed before and after the meeting;
schedule a meeting during the wrong SEASON;
rent MEETING FACILITIES unless there are considerable trade-offs;
hold a meeting in a PLACE which is too expensive for your participants;
hold a meeting in a place with rampant CRIME;
invite a meeting to a politically unstable COUNTRY;
rely on the tourist industry without detailed CONTRACTS;
blindly rely on the local OFFICE OF TOURISM;
allow major SATELLITE SYMPOSIA before your meeting;
publish PROCEEDINGS unless it is worth it;
create unnecessary COMMITTEES;
waste time with meaningless CEREMONIES;
give TRAVEL SUPPORT before the meeting;
accept PERSONAL CHECKS made out in foreign currencies;
accept ABSTRACTS without registration fees;
invite unreliable SPEAKERS;
allow more than two POSTERS per senior author;
schedule SHORT COMMUNICATIONS unless there is a special reason
     to do so;
schedule a CLOSING LECTURE unless it promises to be meaningful;
schedule lengthy AFTER-DINNER TALKS;
set up HEAD TABLES at banquets.

Appendix A

Poster presentations

A frequent mistake made in the design of posters is an overabundance of text. On the other hand, in the information on poster boards, the most serious omission is the lack of precise dimensions of the usable area. The usable area of a poster board will be smaller than the total size if the board extends almost to the floor. Furthermore, the dimensions of poster boards vary greatly and thus may demand a differing design of the title and list of authors' names.

The following example applies to the more user-friendly type of poster board. However, the dimensions of boards may be different (e.g., 1.2 meters (4 feet) wide and 1.5 meters (5 feet) high), which obviously saves space in the exhibition area.

Specific instructions

(1) The poster board (usable area starting 2 feet = about 0.6 meters above ground) is 6 feet (about 1.8 meters) wide and 4 feet (about 1.2 meters) high.

(2) Keep text to a minimum. Use less than a total of 1000 words for the poster (including legends of figures, and tables), and avoid redundancies.

(3) The title should be concise and readable from a distance of at least 7 feet (about 2.1 meters). Use a bold and black typeface, about 1¼ inches (about 30 millimeters) high.

(4) The names of the authors should be somewhat smaller, about 15–20 millimeters high.

(5) The text type should be no less than 5 millimeters for capitals and taller letters such as l or b.

(6) Bear in mind that your illustrations may be inspected from a distance of 1 meter (about 3 feet) or more. Keep them simple and use (but don't overuse) color whenever helpful.

(7) Use figures whenever possible, and avoid complex or unnecessary tables. 'One picture is worth 1000 words.'

(8) Provide a concise abstract (no more than 200 words), a brief introduction and a clear summary with conclusions. Give one or more key references.

(9) If critical for the evaluation of the work, give specific references on the techniques used.

(10) Be creative and use your artistic capability.

(11) Prepare a hand-out (extended abstract with references) and make copies available during the poster session.

Appendix B

Outline of a general schedule for an international meeting

Please see over:

Time	Sunday	Monday	Tuesday	Wednesday	Thursday	Friday	Saturday
9:00–9:45		Opening	Plenary lect.	Plenary lect.		Plenary lect.	Plenary lect.
9:50–10:20		Plenary lect.	SoA lect.	SoA lect.		SoA lect.	SoA lect.
10:25–10:55		SoA lect.	SoA lect.	SoA lect.		SoA lect.	SoA lect.
		Coffee break	Coffee break	Coffee break		Coffee break	Coffee break
11:10–11:40		SoA lect.	SoA lect.	SoA lect.		SoA lect.	SoA lect.
11:45–12:15		SoA lect.	SoA lect.	SoA lect.		SoA lect.	SoA lect.
		Lunch break[b]	Lunch break[c]	Lunch break[d]	All-day excursion	Lunch break[c]	Lunch break
14:00–14:45		Plenary lect.	Plenary lect.	Plenary lect.		Plenary lect.	Plenary lect.
14:45–15:30	Registration[a]	Poster previews	Poster previews	Poster previews		Poster previews	Poster previews

Time	Sunday	Monday	Tuesday	Wednesday	Thursday	Friday	Saturday
15:30–17:00	Registration[a]	Poster discussions	Poster discussions	Poster discussions	All-day excursion	Poster discussions	Poster discussions
17:00–18:40	Registration[a]	Colloquia or Round table conferences	Colloquia or Round table conferences	Colloquia or Round table conferences	All-day excursion	Colloquia or Round table conferences	Business session
		Evening break	Evening break	Evening break		Evening break	Evening break
20:30–23:30	Informal get-together	Welcome party	Workshops	Workshops	Forum?	Workshops	Banquet

[a] Registration will be continued until 22:00; and on Monday and Tuesday 7:30 – 12:00 A.M. Other times will be posted on the bulletin board

[b] International committee luncheon I

[c] Special interest luncheons

[d] Special interest luncheons

[e] International committee luncheon II

Appendix C

Excerpts from a letter to prospective panelists of a Colloquium

(1) The colloquium will begin with an introduction of the panelists by the moderator, followed by a sequence of brief presentations by the moderator and the panelists. Thereafter, the panelists will have the privilege of asking each other questions before the general audience is invited to join in.

(2) The total time allowed for the colloquium is 100 minutes (17 : 00–18 : 40); of which, each panelist will be entitled to 10 minutes of presentation, if he wishes to make one. However, if a panelist prefers to make no formal presentation, he/she will be given an equivalent amount of time for questions to the other panelists immediately following the formal presentations.

(3) The presentations are *not* supposed to be *lengthy ego trips* (though slides and/or an overhead projector may be used, if necessary); rather, they should be a means to make an important point for discussion. Mutual challenges in a congenial atmosphere are encouraged. At the end of the Colloquium, the moderator will summarize the results, i.e., agreements, disagreements, and open or new questions.

Appendix D

Excerpts from a letter to the moderators of Colloquia

Excerpts of the letter referred to in this note appear in Appendix C.

Dear Colleagues,

Though the attached letter to prospective participants of your Colloquium provides most of the essential information, let me emphasize that the success of the Colloquium will depend on the skills of the moderators.

The attached letter will be mailed to all prospective panelists together with the invitation. We hope that its message is clear, and that it will make your chore easier.

As soon as the final composition of your panel of discussants is clear, we will ask you to contact them and to obtain a rough outline of the issues they wish to present. Thereafter, it will be up to you to coordinate the sequence of presentations, and to familiarize all panelists with the issues to be considered.

Hopefully, you will be able to prevent talks that exceed 10 minutes. If the discussants adhere to the time schedule, there will be 40–50 minutes for a general discussion with questions from the audience.

Appendix E

Invitation to participate in a Socratic Workshop

Dear Colleague,

It is my pleasure to invite you to participate in a Socratic Workshop, entitled
, which will be held in a local restaurant on the evening of Tuesday,
May 16, beginning at 8 : 30 pm. Drs XX (UK) and YY (Japan) have kindly agreed
to act as the 'leaders' of the workshop.

The Socratic Workshops are an attempt to break away from the traditional format,
and to open new and better avenues of communication. They have been developed
from many discussions with colleagues who unanimously felt that we need more
time, and a relaxed, informal atmosphere to get to know each other, to ask questions
and to exchange ideas. Since this is a new format, it is unlikely that everything
will be perfect; however, I hope that your participation and cooperation with the
workshop leaders will make it a success that can be built upon in future meetings.

The workshops have been scheduled to last about three hours. They will begin
with an informal cocktail period, followed by a seated dinner (*á la carte*), and
subsequent discussions. The idea of the workshops is to communicate, ask questions
and engage in fruitful interactions. This precludes the use of slide presentations;
the latter would make it too tempting to engage in lengthy monologues, which is
exactly what we do not want. However, flip-charts (blackboard substitutes) will be
available. The workshop leaders have received suggestions concerning the conduct
of the workshops, and they will decide on details of the evening.

Though we cannot make financial commitments at this point, we hope that you
will attend the Symposium. Please let me know by   whether you will be
able to accept this invitation.

Appendix F

Excerpts from a letter to leaders of Socratic Workshops

Dear Colleagues,

Enclosed with this note you will find the preliminary list of prospective participants in your workshop.

Undoubtedly further changes and additions will be requested; in particular, I foresee that some of you may wish to add participants to the list for your Workshop. You are welcome to do so, provided: (*a*) this person has agreed to participate; and (*b*) the person has registered, or will pay the fee for delayed registration upon arrival at the meeting. However, after February 15, it will be too late to add names to the printed program since the latter will then go press.

If questions concerning fees should arise, please make it clear to prospective late registrants that there will be no discount bargaining; in fairness to other participants, we must insist on the full payment for delayed registration.

I would like to emphasize that it is not necessary (and for time's sake not even desirable) that the participants introduce themselves with more than a few remarks on their research interests. After all, they should have submitted their 'statement of research interest' to you, and everybody should have had time to read the 'statements' of the other participants.

When you contact the participants of your Workshop, please include the sample of a 'statement' in your mailing.

If you have further questions, or if you need assistance, please contact me by phone   ('collect,' if necessary), or FAX.

Appendix G

Example of a 'statement of research interests' for a Socratic Workshop

Name of workshop: Molecular biology of avian fright
Participant's name and address: Charles B. Fowler
Department of Molecular Psychiatry
Comb and Wattle University
Chickening, PA 112233
USA
Phone: (717) 999-8888

(1) *Field of research*:
Molecular approaches to stress in broilers.
(2) *Most recent findings*:
Upon second exposure, broilers avoid heated stoves.
(3) *Species recently studied*:
Domestic chicken (*Gallus domesticus*).
(4) *Major techniques recently employed:*
(*a*) Gentle persuasion of old-fashioned chickens.
(*b*) Pacific blotting of resisting roosters.
(5) *Information wanted*:
Impact of barbeque odors on pullets and old boilers.
(6) *Cooperative projects:*
(*a*) Cooperation wanted:
Production of transgenic chickens without olfactory system.
(*b*) Cooperation offered:
Production of slow-moving, deaf and fearless poultry strains.
(7) *Recent publications*:
(*a*) Fowler, C. B. (1999). Limits of genetic engineering: rabbit × chicken crossbreeds grow rabbit ears, but no beak. *J. Incredible Data* **6**: 1–10.
(*b*) Fowler, C. B., Duckett, K., and Splicer, O. (1999). Molecular biology of fear in chickens. *Bandwagon Rev.* **6**: 1–999.

Appendix H

Service contract with an exposition service

The following listing contains points to consider when negotiating with an exposition service. In this example, it is assumed that the executive officer of a scientific society is conducting the negotiations, and that the society: (*a*) has unrestricted control over the exhibition area; (*b*) has prepared an information kit for exhibitors; (*c*) expects considerable interest by exhibitors (e.g., publishers, technological firms); and (*d*) has a list of its regular exhibitors available. Note that it would require considerable effort to organize major exhibitions if your meeting is not of the regular, annual type; and furthermore, that you would need expert advice on the organizational and financial aspects.

(1) The Society will provide an exhibitor service kit, and the Service will provide the necessary forms for inclusion in the exhibitor service kit. These forms have to be approved by the Society. They will contain pertinent information on shipping requirements, instructions for installation and dismantling, and on other available services.

(2) The Service will mail this information to prospective exhibitors upon receipt of a list from the Society, and thereafter whenever requested.

(3) The Service and the Society will jointly develop rules and regulations for the exhibitions.

(4) The Service will provide a draft of the floor plan, tailored to the Society's specific requirements. Upon approval by the Executive Officer of the Society, the Service will prepare a final plan, complete in every detail, and suitable for reproduction.

(5) The Service will provide the labor and decorating equipment to lay out the floor.

(6) The Service will install, maintain and dismantle the required number of exhibit booths. The precise dimensions of these booths must be approved by the Executive Secretary of the Society.

(7) Each booth will include: backwall, side rails, one draped table (size to be approved by the Society), two chairs, one waste basket, and a hand painted exhibitor ID sign (price to be approved by the Society).

(8)   The Service will set up the registration area according to the specifications of the Society, at no charge.

(9)   The Service will receive at their warehouse and deliver to the exhibition site a specified amount of material (material of the Society) at no charge.

(10) The Service carries all required insurance including liability and workman's compensation.

(11) The Service will not charge for overtime.

(12) The Service will provide a manager to be available throughout the exhibition hours.

Appendix I

Contract with a congress hotel

This example of a contract between a large society and a congress hotel is abbreviated and slightly modified from the original. However, numbers of persons and rooms and amounts of money are unchanged. Note that (*a*) the agreement was made four years before the meeting, (*b*) the expected attendance at the meeting is considerable, and (*c*) between Christmas and New Year the guestroom occupancy of many hotels is low, while catered functions are strongly increased. Thus, the hotel commits itself very early to room rates (which usually go up), and also agrees to two parties free of charge (at a time when space for social functions may have brought a profit).

Agreement

Agreement made this 5th day of January, 1995, between The Tombstone Resorts Hotel in Last Rites, Arizona, hereinafter referred to as the Hotel, and the Tobacco Friends Society, hereinafter referred to as the Society.

Purpose of agreement

The agreement specifies the terms under which the Hotel will provide rooms, suites, function spaces, and specified other services for the 1998 Meeting of the Tobacco Friends Society, to be held December 27–30, 1998.

Rooms reserved

(1) The Hotel will hold 1000 rooms available to the Society from December 26 to December 31, 1998. Those rooms not reserved by December 7, 1998 may be made available to persons other than meeting participants; however, participants may continue to book rooms at the special convention rates as long as space is available. It is understood that 700 rooms is a more typical usage for the Society, and 1000 rooms is in anticipation of growth by 1998.

Rates to be charged for rooms and suites; Complimentary rooms

(2) The Hotel will guarantee the rates for rooms at $66 per room for single or double occupancy, and $76 for triple or quadruple occupancy. These rates will be valid from December 23, 1998 to January 3, 1999 for meeting participants who wish to extend their stay. Rates are non-commissionable and are subject to local hotel occupancy tax which is subject to change. The 1995 tax rate is 11%. The Hotel will assign rooms in such a way as to reward those who mail in housing forms early with a view room. The Hotel will provide 50 guest rooms on the concierge level at the above rates with amenities and services normally allocated to this special level. The Hotel will provide 27 executive suites at the above rates, thus making the parlors available without charge.

(3) The Hotel will provide to the Society, free of charge, four VIP suites with two bedrooms each for 5 evenings, December 26–30, 1998. In each suite the parlor and both bedrooms will be complimentary.

(4) The Hotel will provide one complimentary room for every 30 room reservations processed as of December 7, 1998. Occupants of these complimentary rooms have the right to stay up to five (5) nights without payment. The Society will provide the Hotel with a list of names for complimentary accommodation by December 14, 1998.

(5) The Hotel will provide 10 staff rooms at $33 for single or double occupancy and $38 for triple or quadruple occupancy. The Society will provide the Hotel with a list of names for staff rooms and VIP reservations by December 14, 1998.

Reservations

(6) The Hotel will accept reservation cards directly from the meeting participants, reserve the rooms, confirm the reservations in writing within five (5) business days of receipt, and provide a printout of the reservations to the Society's Executive Officer once per week from November 7, 1998 throughout December 21, 1998. The Society will obtain the approval of the Hotel on the Society's housing form before publication, and will bear the cost of publication. Rooms to be held past 6 : 00 pm must be guaranteed with a credit card or a cash deposit of $66 or $76.

Function space

(7) The Hotel will make available to the Society, at no charge, all function space in the Hotel including exhibit space from 8 : 00 am on December 26, 1998 through noon on December 31, 1999. There will be no charge for room setups including, but not limited to, theater style seating, classroom seating, banquet seating, or reception arrangements, nor for changes in setups.

(8) The Society will submit to the Hotel a tentative program 6 months in advance (June 27, 1998) and a final program 2 months in advance (October 26, 1998). The

Hotel will continue to hold all space after the final program, and the Society will release space as the Executive Officer sees fit, keeping in mind the Hotel's need to sell function space that will not interfere with the Society's plans. The Society will hold a majority of its catered events at the Hotel.

(9) The Hotel will confirm with the Society all food and beverage prices 120 days prior to the opening of the meeting (by August 28, 1998). Menu selections will be made by the Society 60 days prior to the meeting (by October 26, 1998).

Miscellaneous

(10) The Hotel will provide on a complimentary basis beer and wine plus hot and cold hors d'œuvres for 200 graduate students for a two-hour party during the meeting. The Hotel will provide a one-hour cocktail reception for 40 attendees (the Society's local committee and invited guests) with hot and cold hors d'oeuvres, mixed drinks, wine and beer. Catering arrangements will be approved by the Society's Executive Officer for both functions.

(11) The Hotel will provide a fully stocked bar setup for two suites on December 16, 1998, 10 VIP food and beverage amenities and 10 roundtrip airport transports by limousine.

(12) The Hotel will consult with the Society before it accepts bookings of other groups at the time of the meeting.

Damages for unavailability of Hotel

(13) The Hotel shall do everything in its power to ensure the availability of rooms in the Hotel. If for unforeseen reasons any part of the room block or meeting space is not available, the Society shall have the option to cancel this contract without financial obligation.

(14) In any controversy between the parties to this Agreement arbitration shall comply with provisions of the State of Arizona. The costs of such arbitration shall be borne as the arbitrator shall decide.

Governing law, entirety of Agreement, and partial invalidity

(15) This Agreement shall be governed by the laws of the State of Arizona. It constitutes the entire agreement between the parties regarding its subject matter. If any provision in this contract is held by any court to be invalid, void or unenforceable, the remaining provisions shall nevertheless continue in full force.

Executed at the Tombstone Resorts Hotel in Last Rites, AZ, on January 20, 1995.

Tobacco Friends Society by: JOHN SMOKE, *Executive Officer*
Tombstone Resorts Hotel by: JOSHUA UNDERTAKER, *Director of Marketing*

Appendix J

Suggestions for the preparation of abstracts and abstract forms for scientific presentations

Abstract forms exist in many varieties. Sometimes, there are almost no instructions; in other cases, there is a verbal overkill that confuses the authors. Occasionally, abstract forms are filled with requests for unnecessary information, and questions that seem to have leaked from the brain of an underburdened bureaucrat. What are important points for the preparation of abstracts?

(1) The format of the abstract. Instead of long explanations, give an example, perhaps as follows:

SCHNAUZER, P.[1], PINSCHER, V.[1] and BOXER, T. R.[2] ([1]Dept. of Barking, Biter College, Dogtown, NJ 08888, USA; [2]Sniffer Inst., Obedience Academy, Dogwood, L23 PP3, England). *Dog tags: mark of distinction or insult?*
'Heightened awareness' is confusing human (*Homo sapiens*) minds. Dogs begin to wonder . . . Conclusion: intelligent dogs should smile and take their tags as symbols of human simplicity. (Supported by grant # BITE 2 from NDA). *Ref.*: Poodle *et al.* (1994), *Dogs Daily* **8**: 23–44; Schnauzer & Sniffer (1995), *J. Fido Res.* **16**: 45–88.

(2) A reminder that abstracts will be rejected if they exceed the allotted space, or a total number of 200 words of text (excluding title, addresses and citations). Emphasize that it is important to practise before typing the final version.

Especially at international meetings, a surprising number of people seem to be blind to printer's cut lines. On the other hand, it is easy to miss the rectangle when a properly prepared abstract is transferred from a word processor. This is no problem, provided the rectangle is given in heavy dark lines on a light background as a typing guide, and abstracts are accepted on plain white paper. However, when specially designed forms must be used and authors receive only a single copy, mistakes can lead to expensive telephone calls and other problems.

(3) Precise information on the fonts to be used (e.g., TIMES, COURIER or

ELITE 12). This is very critical if the abstract is to be reduced during the printing process.

(4) Information on hand printing of letters and symbols that are not on the typewriter or computer. Usually, black ink is required.

(5) Information on inclusion of figures and tables.

(6) A reminder that abstracts must NOT be folded, cut out, stapled or taped; and that the complete page should be mailed with a piece of cardboard of the same size (fitting snugly into the envelope) to avoid damage.

(7) Information on the number and selection of key words that will be needed for the index of the abstract volume.

(8) Deadline for receipt of the abstract.

(9) Exact address (including telephone plus e-mail or FAX number) of the person/ office to which the abstract must be mailed.

(10) Signature of the sponsor if the abstract is submitted by non-member of a society.

(11) Mailing address of the first author, plus telephone, e-mail or FAX number.

Final note Especially at major meetings, it is advisable to mail abstract forms together with a postcard on which the senior author of the abstract fills in his address, and the title of the abstract. This postcard will be sent back by the editor to the senior author, confirming receipt of the abstract. In the case of poor postal services, this procedure will cut down on telephone calls by concerned authors.

Appendix K

Letter *with clout* to a manuscript delinquent

Dear Colleague,

I hate this letter as much as you will certainly do. However, being responsible for the publication of the proceedings of the symposium, I have 'to bite the bullet.'

As you know from our recent telephone conversations, we are almost ready to hand in the more than 100 manuscripts of the symposium talks to the publisher; and at this moment, there are only a few contributions outstanding. However, without these remaining manuscripts, my coeditors and I are unable to prepare the index for the approx. 900 pages of this volume.

Please, understand that several hundred people from more than 40 nations have already paid $85 for the proceedings, and that we don't need an avalanche of justifiably nasty letters from all over the world. Also, please recall that we have a signed agreement with the publisher which, when broken by us, gives him the liberty to postpone the publication until convenient for him; and if the volumes do not appear this year, they will be almost obsolete. Finally, I must mention that I promised the National Science Foundation the publication of the proceedings in 1990. This promise was certainly considered when they decided on the travel grant award, from which you will receive a total of $600. I do not wish to withhold travel funds from manuscript delinquents; however, as you know, I will have to justify the distribution of the funds in the final report, and I have a tendency to be brutally frank and outspoken.

At any rate, instead of wasting your and my time with additional and moral and other arguments, let me remind you that I volunteered to organize your workshop for the symposium; this was an added burden which took several days of working time when the pressures of the final preparations of the symposium were almost unbearable. Therefore, I hope that you can take now, almost six weeks after the deadline, a few hours (and that is all it should take) to write the manuscript immediately.

Appendix L

Schedule for staff of an international meeting

Note It is assumed that: Dr Jones is the organizer of the meeting; Dr Fernandez is coeditor of the proceedings. Mrs Smith has been involved in the preparations of the meeting and is thoroughly familiar with details; she is now in charge of the registration desk and related organizational details. Drs Black (chairman), Brown, Green and White are members of the local committee. Dr Black and his local committee are cooperative, and they have recruited assistants (graduate students).

The arrival times of most participants at the airport are known and reflected in the schedule. Shuttle buses have been hired for the peak hours.

The following examples are modified excerpts from a staff schedule of an international symposium.

I. During the week before the symposium

(A)   Check the bank accounts so that payments to hotels and other parties can be made (Drs Fernandez and Jones, Mrs Smith).

(B)   Alert the hotels to have sufficient change for the symposium participants (Mrs Smith).

(C)   Alert the hotel receptions to have information on car rentals, theaters, casinos and golf courses (Dr Black).

(D)   Arrange with the congress hotel details of the welcome party (Drs Black and Jones, Mrs Smith).

(E)   Arrange details of the delivery of projectors, flip-charts and other equipment with supplier (Drs Black and White).

(F)   Provide for the poster areas: (1) push pins (9000); (2) heavy tape; (3) scissors; (4) individual numbers 1–1000 (on paper or cardboard 10 × 10 centimeters) for poster boards (Dr White and one assistant).

(G)   Prepare 600 lists of participants (with addresses), to be included in the portfolios (Mrs Smith and two assistants).

(H) Buy envelopes for tickets, refunds, financial support etc. (Mrs Smith and one assistant).

(I) Get memo pads and sticky tape for notes to be attached to the portfolios (Mrs Smith and one assistant).

(J) Provide for the registration desk: computer system with spare diskettes and printout paper; calculators (at least one with printout); typewriter with large typeface (for name tags); change (2000 US dollars); forms (receipts; confirmation of attendance, etc.); boxes for manuscripts and money; writing utensils; two staplers with extra staples; paper clips; folders; note book and hole punch (Drs Black, White and Fernandez, Mrs Smith and two assistants).

II. Sunday, May 14, 1989

(A) *Morning*:   *Time*

  (1) Prepare portfolios: 11 : 00–13 : 00
     (Mrs Smith and two assistants)

  (2) Set up message/bulletin board (with 1000   12 : 00
     thumb tacks in attached paper cups)
     (Drs Black and White, staff of the hotel)

(B) *Afternoon*:

  (1) Set up signs for meeting rooms: 14 : 15–14 : 30
     (Drs Black and White; staff of the hotel)

  (2) Set up symposium office:

    (a) Check setup for symposium office,   14 : 30–14 : 45
     including sign with the opening hours:
     (Drs Black and Jones, Mrs Smith; staff of
     the hotel)

    (b) Briefing of the staff of the symposium   15 : 00–15 : 15
     office:
     (Drs Black and Jones, Mrs Smith; three
     assistants of the symposium office)

    (c) Final preparations of symposium   15 : 15–15 : 30
     office:
     (Drs Black and Jones, Mrs Smith; three
     assistants of the symposium office)

  (3) Open hours of the registration desk: 16 : 00–22 : 00
     (Drs Fernandez and Jones hourly alternat-
     ing for collection of manuscripts; Mrs
     Smith and three assistants all the time)

  (4) Delivery and check of equipment: 16 : 00–17 : 00
     (Drs Black and White, four projectionists,

two trouble shooters; supplier of projection equipment; staff of the hotel)

(5) Inspect poster areas and attach number   17 : 00–17 : 30
signs to the boards:
(Drs Black and White, two trouble
shooters)

(6) Welcome and assistance for the arriving
participants.
(*a*) At the airport: 15 : 00–22 : 00
(*b*) At the congress hotel (entrance):   16 : 00–23 : 00
(Drs Fernandez and Jones hourly alternat-
ing, two assistants)

III. Monday, May 15, 1989

(*A*) *Morning*:   *Time*

(1) Symposium office: 08 : 00–12 : 00
Open hours of the symposium office: (Drs
Fernandez and Jones hourly alternating
for collection of manuscripts; Mrs Smith
and three assistants all the time)

(2) Meeting rooms
(*a*) Equipment check: 08 : 15–08 : 45
(Drs Black and White, two trouble
shooters)
(*b*) Two guards checking the name tags: 08 : 45–12 : 00
(*c*) One projectionist and one trouble   08 : 45–09 : 45
shooter:
(*d*) Four projectionists and two trouble   09 : 30–12 : 15
shooters:

(3) Accompanists' programs:
(*a*) Assistance at the departure of the   09 : 30–09 : 45
buses:
(Drs Black and White, one assistant)

(4) Welcome and assistance for arriving par-   09 : 00–15 : 00
ticipants at the airport:
(Drs Brown and Green, three hours each;
two assistants)

(*B*) *Afternoon*:

(1) Registration desk:
Noon shift: 12 : 00–12 : 30
(Dr Black and two assistants)

Afternoon/evening shift:	12 : 30–19 : 00
(Mrs Smith, one assistant)	

(2) Poster area:

(*a*) Assistance in setting up of posters:	12 : 00–14 : 30
(Dr White, one trouble shooter)	
(*b*) Further assistance:	14 : 30–17 : 00
(Dr Black, one trouble shooter)	

(3) Meeting rooms:

(*a*) Two guards checking the name tags	13 : 45–18 : 15
(*b*) One projectionist and one trouble shooter	13 : 45–14 : 45
(*c*) Five projectionists, one trouble shooter	16 : 45–18 : 45
(Drs Fernandez and Jones sitting in on Colloquia)	

(4) Evening reception: 20 : 30–23 : 30
     Two guards checking name tags

Appendix M

Checklist of equipment and supplies for the registration desk of a major meeting

(1) Bulletin/message board and mail board.

(2) Sign indicating location of the registration desk.

(3) Desk and chairs.

(4) Completed and coded ('problem' or 'message') portfolio sets for registered participants. These sets may include: list of registered participants; note pad and pen; name tag and holder; tickets; refunds and/or financial support (with receipts to be signed); updated meeting information; local information and/or advertisements (with coupons for discounts).

(5) Portfolio sets for late registrants, containing the same as in (4), but also program and abstracts. Tickets may have to be added.

(6) Calculators *with printout paper* (important for instant check of calculations!).

(7) Computer with diskettes, printer, printer paper. If there are security problems, use a laptop (notebook) type computer that can be taken easily to a safe place.

(8) Typewriter with large typefaces for name tags.

(9) Pencils, pens and markers.

(10) Note pads and other paper, folders and envelopes.

(11) Sticky tape, scissors and thumb tacks (for boards).

(12) Stapler with extra staples, paper clips, hole punch.

(13) Receipts and other forms.

(14) Tickets for various events.

(15) Cash box for money and change.

(16) Copies of the list of participants.

(17) Copies of program and abstract volume.

(18) Maps and local information, including telephone numbers of taxis, ambulance and police.

(19) Copies of hotel reservations for participants.

(20) Telephone.

(21) Spare frames for slides.

(22) Spare foils and markers for overhead projections.
(23) Address list of all members of the Society.
(24) Mobile phone (cell phone).

Appendix N

Equipment and supplies for meeting rooms

Example of a checklist for the first day of a major meeting with parallel sessions.

Monday, May 15, 1989

(A) Morning
(1) Four slide projectors for use, two spares; four extension cords; two spare bulbs.
(2) Four supports for projectors.
(3) Four wireless pointers, two spares.
(4) Four screens of adequate size (check rooms).
(5) Four wireless microphones, two spares.
(6) Four podiums with lights (plus spare bulbs).
(7) Reserved seats (front row) for speakers and special guests of the opening ceremony.

(B) Afternoon
(1) Five slide projectors for use, one spare; five extension cords; two spare bulbs.
(2) Five supports for projectors.
(3) Five pointers for use, one spare.
(4) Five screens (check fifth screen for adequate size).
(5) Two overhead projectors (for Colloquia).
(6) One hundred poster boards with running numbers in upper right corner.
(7) 9000 push pins for posters.
(8) Five wireless microphones, one spare.
(9) Five podiums with lights (plus spare bulbs).

(C) Evening
Four flip-charts and markers.

Appendix O

Example of a letter/form confirming attendance of prospective participants in various events of a meeting

The response to this letter will help identify participants that have to be replaced before the program is finalized. To avoid misunderstandings, the events (Session, Workshop, Colloquium) should be specified in the form.

Dear Colleague,

During the preparations of the program for this symposium, we have aimed to achieve a maximum of participation and a minimum of overlaps. This required endless hours of scheduling and rescheduling of several hundred time slots.

Since we are trying to avoid unnecessary, time-consuming and expensive paper-work, we ask you to fill out the bottom of this form and *return it, together with the green Registration Form*, to .... as soon as possible (address on top of this form), but no later than June 15, 1989. We will take the return of the filled form as an indication that you are definitely planning to attend the symposium, regardless of whether travel support will be available or not.

(1) I accept the invitation to participate in the following event(s):

(*a*) State-of-the-Art Lecture, entitled 'Prognosis of egg production in anencephalic chickens'

_____

(*b*) Participation in workshop 3 on 'Avian ovulation'

_____

(2) I wish to modify the title of my talk as follows:

_____

_____

Signature and Date: _____

Appendix P

Calculation of fees

Example of information on registration fees, including authorization for charges to credit cards. Note that the amounts are *double-checked*. Use the same approach for calculating fees for social events and other expenditures.

REGISTRATION FEES

	Until May 16, '00	After May 16th, '00		Number of persons		Total
Member of society	$100	$120	×	_____	=	$_____
Non-member	$120	$150	×	_____	=	$_____
Graduate student	$60	$70	×	_____	=	$_____
Accompanists	$60	$70	×	_____	=	$_____

TOTAL REGISTRATION FEES: $_____

Note

(1) Lower fees must be *received* (*not* postmarked) by May 16.
(2) Registration forms received without correct payment will not constitute a valid registration.
(3) Foreign payments in *US dollars only* by bank draft (bank check), electronic mail or credit card. Only Mistercard, Vista and Card Beige will be accepted; for instructions, see below.
(4) Nationals and permanent residents of Lavachia pay in local currency (see enclosed note).
(5) *No personal checks* will be accepted.

Payments to: TICSL Account, National Bank of Lavachia, Avenue of the Great Leader, 12345 Bongout, Lavachia.

Cancellations:

100% refund if withdrawal notice is received before December 1, 2000; 80% thereafter. No refunds after March 21, 2001.

****

Authorization

For *payments with credit card*, detach this section and mail it to:

TICSL, Faculty of Science, University of Chuleta, 56789 Chuleta, Lavachia.

I herewith authorize TICSL to charge the amount of US $ _____

(print amount):

to the following credit card:

Number:

Expiration date: Day _____ Month _____ Year: _____

Name (printed):

             FAMILY NAME   First name   Middle initial

Name (signed): _____   _____   _____

Date signed: _____

Appendix Q

Suggestions for a hotel reservation form

Since the information provided by hotels and the local office of tourism often lacks sufficient detail, use the following form as a checklist.

Note: (1) There should be at least five lines for the address since many academic institutes have excessively long addresses. (2) There is an evergreen confusion with phone and fax numbers. In many countries, the local (area) code is preceded by a '0' or '9' if calls are national; this additional number must be omitted for calls from a foreign country. (3) The names of some meetings are very long, requiring two lines of text, while in other cases simple acronyms can be used.

If you find important items missing in the official forms of your venue, insist that they include the missing details in the mailing to your participants. It may cut down on ill feelings and complaints about your poor choice of a meeting site. To the following, add the authorization given in Appendix P.

Lodging reservation form
(please, print)

Your name _____

        FAMILY NAME   First name   Middle initial
        (capitalize)

Name of your meeting _____

Your address _____

_____

_____

_____

_____

Phone: _____

      (national calls)   (international calls)

Fax: _____

     (national)   (international)

E-mail: _____

(A) Lodging request for hotel

Date of arrival: March   , 2001
Date of departure: March   , 2001. Total number of nights:
Number of adult persons in party:
Number of children aged 12 or younger:

Note: All hotel rooms have single or twin beds. Children aged 12 or younger will be accommodated on a fold-up bed, free of charge, in their parents' room. All rooms have air conditioning and private bathrooms.

Preferred hotel:		First choice	Second choice
OCEANVIEW: (four-star)	$70 per night single occupancy	_____	_____
	$60 per person per night (double occupancy)	_____	_____
LAKESIDE (three-star)	$60 per night single occupancy	_____	_____
	$50 per person per night (double occupancy)	_____	_____
TWO RIVERS (three-star) Old Wing	$55 per night single occupancy	_____	_____
	$45 per person per night (double occupancy)	_____	_____
New Wing	$60 per night single occupancy	_____	_____
	$50 per person per night (double occupancy)	_____	_____

NOTE: These prices will increase by $10 per night (per person) for all reservations arriving after January 1, 2001.

If you wish to share a room and have *separate* bills:

I wish to share my room with

(Print full name of person; capitalize FAMILY name)

(Address)

(B) Lodging request for the International House (student residence)

(Four persons per room; $20 per person per night)

Date of arrival: <u>March   , 2001</u>

Date of departure: <u>March   , 2001</u> Total number of nights: _____

Deposit
All hotels require a deposit of $100 per room. However, accommodation cannot be guaranteed after March 1, 2001.

Cancellations
100% refund if withdrawal notice is received before December 1, 2000; 80% thereafter. No refunds after March 21, 2001.

The student residence requires a non-refundable deposit of $20 by December 1, 2000.

All efforts will be made to honor your first choice. However, rooms will be given on a 'first come first served' basis.

Only a limited number of single rooms are available.

Send your *deposit* with this form to: LAVACHIA INTERTOURIST, Avenue of the Great Leader, 12345 Bongout, Lavachia.

Appendix R

Summary of daily program schedule

Example using the schedule for Tuesday of Appendix B. Note that: (*a*) the parallel morning sessions avoid overlap of topics; (*b*) the topic 'egg production' had to be split into two sessions because of the large number of pertinent papers (Session B will be held on another day); (*c*) Poster Sessions and Colloquia relate to the topics of the morning sessions; (*d*) the Socratic Workshops deal with topics not specifically considered in the other events of the day (to avoid too much discussion of the same matter on one day, and to offer variety for persons with different interests).

Tuesday, March 21, 2001

Morning/early afternoon
9 : 00–9 : 45 Plenary Lecture
    The cholesterol crisis
9 : 50–12 : 15 Parallel sessions
    (1) Egg production A
    (2) Shark grooming
    (3) Hoof disease
    (4) Goat milk

12 : 15–14 : 45 Lunch break

14 : 45–17 : 00 Plenary Lecture
    The calcium craze
14 : 45–17 : 00 Poster Sessions
    (1) Cholesterol metabolism
    (2) Avian reproduction
    (3) Fish farming A
    (4) Cattle diseases
    (5) Milk consumption
    (6) Calcium metabolism

Late afternoon/evening
17 : 00–18 : 40 Colloquia
    (1) Cholesterol: Friend or foe?
    (2) Eggs of transgenic ostriches
    (3) Sharks in Utah lakes
    (4) Buffalo footwear
    (5) Sour milk marketing

18 : 40–20 : 30 Evening break

20 : 30–23 : 30 Socratic Workshops
    (1) Green beer effects in Kobe cows
    (2) Stress and odor in chickens
    (3) Crayfish farming in bathtubs
    (4) Canary colors
    (5) Two-tailed cat races
    (6) Standards for salami sausage
    (7) Dog meat: Still a tabu?
    (8) Hog hygiene

Index